Tunnel lining design guide

The British Tunnelling Society and
The Institution of Civil Engineers

Institution of Civil Engineers

D0770298

Thomas Telford

Published by Thomas Telford Publishing, Thomas Telford Ltd,
1 Heron Quay, London E14 4JD.

URL: http://www.thomastelford.com

Distributors for Thomas Telford Books are
USA: ASCE Press, 1801 Alexander Bell Drive, Reston, VA 20191-4400, USA
Japan: Maruzen Co. Ltd, Book Department, 3–10 Nihonbashi 2-chome, Chuo-ku, Tokyo 103
Australia: DA Books and Journals, 648 Whitehorse Road, Mitcham 3132, Victoria

First published 2004

Also available from Thomas Telford Books
Specification for Tunnelling. The British Tunnelling Society and The Institution of Civil Engineers. ISBN 07277 2865 2
Building Response to Tunnelling. The Construction Industry Research and Information Association and Imperial College, London. ISBN 07277 3117 7

A catalogue record for this book is available from the British Library

ISBN: 0 7277 2986 1

Typeset by Academic + Technical, Bristol
Printed and bound in Great Britain by MPG Books Limited, Bodmin, Cornwall

Dedication

David Wallis
1941–2000

Chairman of the British Tunnelling Society
1999–2000

This publication is dedicated to the memory of David Wallis in that the Guide was one of several projects driven forward by him during his chairmanship of the British Tunnelling Society (BTS), cut short by his untimely death in November 2000. The background to the concept of the Guide is given in the Foreword, which was first drafted by David.

It was originally hoped that the Guide would be published at the end of his normal chairmanship period in October 2001. However, the work pressures placed on many members of the working group during a period of, fortunately, increasing tunnel design activity, but limited availability of experienced tunnel engineers, has unavoidably delayed its appearance. A contributory factor was the determination of the working group to carry forward David Wallis' insistence on 'getting it right', as engineers are expected to do, and to provide practical recommendations and guidance rather than less focussed theory.

Contents

Foreword

The need for a single reference of recommendations and guidance for tunnel lining design has been recognised for a number of years, as evidence by discussions in the pages of tunnelling industry journals, at conferences and at the meetings of bodies such as the British Tunnelling Society. Hitherto, designers have adopted a variety of approaches based on practical experience of tunnels built in similar circumstances and on research carried out both on mathematical and scale physical models, either undertaken by themselves or which have been presented in published papers. Combined with such existing knowledge, existing codes and standards, which have not been specifically written for, or appropriate to, tunnelling have been modified.

The need for, perhaps more uniform, tunnel design guidance was precipitated by some well-publicised tunnel collapses during construction, and by the ever increasing demands on tunnelling engineers to increase the parameters within which secure underground excavations could be made, whilst maintaining a competitive stance against other possible solutions to problems in transport, utilities, storage and society's similar needs.

Tunnels are almost unique structures in that they are surrounded by ground of many different types and this has a direct relationship to the type and degree of tunnel supporting lining required. The ground may even be enlisted to aid support of the excavation. In this context, the development of tunnel lining design has included special consideration of such issues as the interaction between the lining and ground, the relatively high compressive loading in relation to bending, the application of loading to structural elements before materials reach maturity, and many others where existing orthodox construction design recommendations are inappropriate.

The British Tunnelling Society (BTS) considered that the valuable knowledge and experience of its members on tunnel lining design should be made available to the wider international underground construction community, and that a published guide was an appropriate medium. A letter to the Editor of *Tunnels & Tunnelling International* in October 1998 finally prompted action by the then Chairman of the Society. Funding for production of the Guide was sought and provided equally by the BTS and the Institution of Civil Engineers Research and Development Fund.

The Guide is drafted for particular use in conjunction with relevant United Kingdom Standards, Codes of Practice, customs and practice (see Bibliography and section references). Such existing Standards and Codes are usually not specific to tunnelling, and have no formal standing in tunnel lining design, so this document carries new information and guidance. Best practice from elsewhere in the world is recognised and adopted where appropriate, but no attempt has been made to comply with any associated norms.

The authors trust that they have met most of the current needs of tunnel designers with the following, but will welcome comments and suggested improvements. These should be sent to the BTS Secretary at the Institution of Civil Engineers, One Great George Street, London SW1P 3AA, England; telephone (+44) (0)207 665 2233; fax (+44) (0)207 799 1325; E-mail: bts@ice.org.uk.

Acknowledgements

The production of the *Tunnel Lining Design Guide* (Guide) was made possible by equal funding from the financial resources of the British Tunnelling Society (BTS), and the Research and Development Fund of the Institution of Civil Engineers.

The BTS is grateful to all those, authors and reviewers, who have given of their time freely despite, in most cases, great pressures on their time from other work. All work on the Guide, apart from specialist editing and publishing services, was unpaid.

Members of the working group
Chris Smith (Chairman)
Maurice Jones (Editor and secretariat)

John Anderson
Malcolm Chappell
John Curtis
Peter Jewell
Steve Macklin
Barry New
David Powell
Steve Smith
Alun Thomas

Other major contributions by
Lesley Parker
Paul Trafford

Produced with additional funding from the Institution of Civil Engineers Research and Development Fund

1 Introduction

1.1 Scope

This Guide is intended to cover the design of structural linings for all manner of driven tunnels and shafts to be constructed in most types of ground conditions. A bibliography is provided of source data and references for more detailed understanding and analysis, and for use where hybrid designs do not fit one particular category described in this guide.

1.2 Background

The need for a single reference of recommendations and guidance for tunnel lining design has been recognised for a number of years. Hitherto, designers have adopted a variety of approaches based on practical experience of tunnels built in similar circumstances and on research carried out both on mathematical and physical scale-models, either undertaken by the designers themselves or which have been presented in published papers. Combined with such knowledge, existing codes and standards that have not been specifically written for, or are not appropriate to, tunnelling have been modified.

Engineers designing and constructing tunnel lining support systems are responsible for ensuring that the selected information provided in this Guide is appropriate for particular projects and for adjusting such information to the particular circumstances of the project. In particular the reader's attention is drawn to those sections of this Guide dealing with risk management and quality control.

In the development of tunnel lining design special consideration has been given to such issues as the interaction between the lining and ground, the relatively high compressive loading in relation to bending, the application of loading to structural elements before materials reach maturity, and many other issues where existing orthodox construction design recommendations are inappropriate.

The International Tunnelling Association (ITA) has had the subject on its agenda for a number of years. This has resulted in the publication of guidelines (International Tunnelling Association, 2000) for the design of shield tunnel linings. This Guide indicates where any differences in recommendations occur.

1.3 Guide structure and objectives

This Guide is primarily intended to provide those determining the required specification of tunnel linings with a single reference as to the recommended rules and practices to apply in their design. In addition, however, it provides those requiring to procure, operate or maintain tunnels, or those seeking to acquire data for use in their design, with details of those factors which influence correct design such as end use, construction practice and environmental influences.

Separate sections are provided following, as far as possible, the sequence of the design process.

1.3.1 Chapter 2 – Project definition

The client has to provide details of required operating and serviceability requirements including design life and maintenance regime, as well as environmental constraints that may be imposed by the location of the tunnel and by adjacent structures that may be sensitive to settlement, or noise and vibration.

Whole life costing will be affected by the client's attitude to the balance between capital and operating costs, lowest cost or certainty of out-turns, as well as the required design life. Clients unfamiliar with modern tunnelling may well require advice on how best to achieve these objectives (Muir Wood, 2000).

The philosophy for structure and construction safety, and risk management, is governed by legislation. Safety aspects are considered in the context of European legislation but have worldwide relevance. All parties need to be aware and anticipate the requirements of evolving standards of safety, especially for road tunnels.

Project financial risk management needs to be defined, including the development of risk sharing and the role of quality assurance and control.

1.3.2 Chapter 3 – Geotechnical characterisation

The process of desk study, field investigation and testing is described, reflecting the means of classifying defined soils and rocks. A clear distinction is made between 'soft' and 'hard' ground. The interpretation of geotechnical data and derivation of design parameters, their range and uncertainty, is explained. The importance of summarising data in a Geotechnical Baseline Report is emphasised.

1.3.3 Chapter 4 – Design life and durability

This chapter reviews the durability requirements of a tunnel, based on its use, and those durability considerations that are dependent on the type of lining system chosen. The effects of different ground and environmental conditions are considered, as well as the effects of various lining types on the durability. The effects of fire are also considered and the various methods of control are examined.

1.3.4 Chapter 5 – Design considerations

This chapter follows through the design process examining failure mechanisms, time dependent behaviour and control of deformations. The selection of an appropriate design approach is outlined together with the application of load cases, and the conditions that influence design are considered.

Available lining systems, together with the basis of selection, and detailed considerations such as tolerances, durability, and watertightness are examined.

While this Guide is not intended to recommend specific construction methods, nor temporary ground support, it is vitally important to take them into account when establishing a lining design. For successful tunnelling, the methods of construction are highly interrelated with the design and other elements of the project. Methods of excavation and control of ground movement are reviewed together with the influences of other conditions, such as groundwater control. Special considerations for the design and construction of junctions, portals and shafts are covered.

1.3.5 Chapter 6 – Theoretical methods of analysis

This chapter deals with the methods of structural analysis, and the derivation of the effective dimensions required. The validity of the various theoretical methods of design is explained and guidance is given on the use of these methods under different conditions.

1.3.6 Chapter 7 – Settlement

In determining appropriate ground support, it is necessary to be able to predict ground movement and its effect. Methods of analysis are explained together with the assessment of the effect of ground movement on adjacent structures. Means of mitigation through lining design and other means such as compensation grouting are described. This chapter also considers the influences of construction on settlement and measures that may be taken to mitigate its effects.

1.3.7 Chapter 8 – Instrumentation and monitoring

Guidelines are given for ground and lining monitoring appropriate to different support considerations, and recommendations are made for the instruments themselves and the capture, storage, interpretation and reporting of data.

1.3.8 Chapter 9 – Quality management

This chapter examines the application of quality systems to design process and installation, whether the materials are prefabricated or formed on site. It is essential to ensure that the designer's intent is achieved within the assumed design allowances, and that deviations are detected and timely remedial action taken.

1.3.9 Chapter 10 – Case histories

The final chapter includes four case histories from recent major projects; three of them give a brief outline of the design process and the parameters used in each case whilst the fourth concentrates on the monitoring arrangements for a particular tunnel. The contracts covered are the Heathrow Express Station Tunnels at Terminal 4; Channel Tunnel; Great Belt Railway Tunnels; and the North Downs Tunnel on CTRL.

1.4 Definitions

There is a wide range of terms used in the tunnelling industry, many of them appear to be interchangeable, and a number are often used synonymously (see definition of 'Support systems', 1.4.1). Definitions of tunnelling terms as detailed in BS 6100: subsection 2.2.3 : 1990 shall apply unless stated as follows.

- **Design** Is taken to mean, for the purpose of tunnel lining construction, the complete process (see 'engineering design process' below) of specifying the tunnel lining requirements. This includes the establishment of project end-use requirements, defining ground and material properties, analysing and calculating structural requirements, identifying construction assumptions and requirements, and detailing inspection and testing regimes.
- **Driven tunnel** Is taken to mean any underground space constructed by enclosed methods and where ground support is erected at or near the advancing face (rather than cut and cover, immersed tube, jacked pipe or directionally drilled methods).
- **Engineering design process** Refers to all design-related activities from concept through to the post-construction stage.
- **Hard ground** Is ground comprising rock that, following excavation in a tunnel face and the removal of or support to any

loosened or unstable material, would be expected to remain stable for an extended period.

- **Hazard** Is something with the potential to lead to an unfavourable outcome or circumstance for the project, or for anybody or anything associated with it. By definition, any hazard that is not identified is uncontrolled and measures cannot be taken to mitigate any potential risks (see definition of 'risk' below).
- **Lining** Is taken to mean the necessary permanent ground support system to the periphery of a tunnel or shaft excavation, and/or the material installed in the same position with an inner surface suitable for the specific end-use of the underground excavation. The lining may vary from limited support in a stable rock formation to continuous support in unstable ground. This publication offers guidance on the design of permanent linings rather than any temporary support used during the construction period, save where temporary support may also be considered to be part of the permanent lining. Therefore, the term 'lining' does not normally include temporary support. See also definitions for 'Support systems' in 1.4.1.
- **One-pass lining** A system of support that is installed integrally with the advancing heading. This could include segmental linings or several layers of reinforced or unreinforced shotcrete applied tight up against the advancing heading.
- **Risk** Is the likelihood of a particular hazard being realised together with the consequences for persons should that occur.
- **Risk management** Is the process of identifying, analysing, assessing and controlling risks on a project. Also known by the acronym 'RAM' from Risk Analysis and Management.
- **Shaft** Is taken to mean a vertical or subvertical excavation of limited cross-section in relation to its depth in which ground support is provided as excavation proceeds, (rather than installed in advance from the surface such as the case of piling or diaphragm walls).
- **Soft ground** Is any type of ground that is not to be relied upon to remain stable in the short, medium or long term following its excavation in a tunnel face.
- **Volume loss** Or 'ground loss' into the tunnel is usually equated to the volume of the surface settlement trough per linear metre expressed as a percentage of the theoretical excavated volume per linear metre.

1.4.1 Support systems

Support terms are often used synonymously, for example temporary and primary support or permanent and secondary support. In the past, this suited the industry contractually, the contractor was responsible for the design of temporary support and the designer for the design of the permanent works, but the position has changed in recent years (see Section 5.2). Support is divided into primary, permanent and temporary support as follows.

- **Primary support system** All support installed to achieve a stable opening is primary support. This will be specified by the designer and may or may not form part of the permanent support system.
- **Permanent support system** Support elements that are designed to carry the long-term loads predicted for the lining system. It may be a design requirement that part or all of the primary

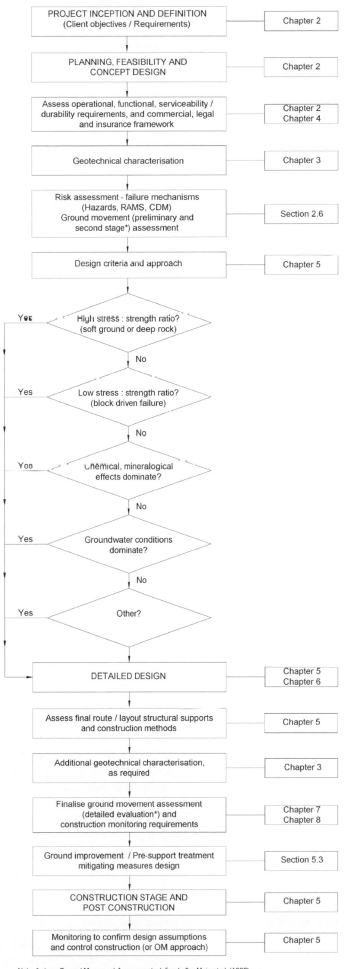

Fig. 1.1 Design Guide sequence

Note: * stage Ground Movement Assessments defined after Mair *et al.* (1996).

system is incorporated into the permanent lining. Whether primary shotcrete is included depends on the specification of the material. In this context, the often quoted assumption that primary shotcrete degrades to a gravel should be avoided in specifications. However, it can be made clear in design briefs that a mix is deliberately designed not to be durable in the long term, and that any lining formed from this material cannot form part of the permanent support.

- **Temporary support** Support that is installed only for temporary purposes, for example internal propping of a segmental lining, spiling, canopy tubes and bolts installed in a heading to improve face stability but that do not form part of the permanent support system.

1.5 Design process

In planning the approach to design it is useful to look in relatively simple terms at the stages involved before developing complex flow charts related to specific activities. Typically, most projects pass through concept, detailed design, construction and post-construction stages. Of these, the concept stage is the most critical in that the entire engineering design process is driven by the decisions made at this stage. It also directs the organisation and lines of communication for the project and is essentially a planning stage in which the most important underlying principle should be that risk management is not optional.

This Guide looks at the various stages in the design process and examines the critical areas; the sequence of the Guide follows the design method as can be seen in Fig. 1.1. The Guide considers the concept, final usage requirements, geological constraints, detailed design and design methods in choosing the type of lining. There are a wide variety of lining systems available and the design approach adopted will ultimately be influenced by the choice of construction method.

Areas of concern with tunnels are also highlighted; stability problems in tunnels are unacceptable, particularly if they could lead to loss of life. There are many factors that can contribute to concerns over stability; for example unforeseen geological conditions (Sections 3.5 and 3.7), poor appreciation of the need to control deformations, late installation of support because of a lack of familiarity with the design basis and poor appreciation of the mechanical limitations of the support system and any lining repair or alteration.

It is increasingly the case that such situations are controlled by improved design and risk management procedures that ensure continuity from design through to construction. Reports such as those prepared by the Institution of Civil Engineers (ICE) (1996) and the Health and Safety Executive (HSE) (1996) in response to the collapse of tunnels at Heathrow Airport in 1994 partially reflect this process. Muir Wood also covers the management of the design process in his publication *Tunnelling: Management by Design* (2000).

1.6 References

Health and Safety Executive (1996). *Safety of New Austrian Tunnelling Method (NATM) Tunnels*. HSE Books, Sudbury, Suffolk.

Institution of Civil Engineers (1996). *Sprayed Concrete Linings (NATM) for Tunnels in Soft Ground*. Institution of Civil Engineers design and practice guides. Thomas Telford, London.

International Tunnelling Association (2000). Guidelines for the Design of Shield Tunnel Lining (Official Report Work Group No. 2). *Tunnelling & Underground Space Technology* **15**(3), 303–331. Elsevier Science, Oxford.

Mair, R. J., Taylor, R. N. and Burland, J. B. (1996). Prediction of ground movements and assessment of building damage due to bored tunnelling. *Proc. Int. Symp. on Geotechnical Aspects of Underground Construction in Soft Ground* (eds Mair, R. J. and Taylor, R. N.). Balkema, Rotterdam, pp. 713–718.

Muir Wood, A. M. (2000). *Tunnelling: Management by Design*. E & F N Spon, London and New York.

2 Project definition

2.1.1 Purposes

Permanent linings are required in many tunnels for two purposes:

- **Structural** To support the exposed ground thus providing and maintaining the required operational cross-section and, if required, to provide a barrier to the passage of liquids.
- **Operational** To provide an internal surface and environment appropriate to the functions of the tunnel.

2.1.2 Construction

The chosen lining must be capable of safe and economic construction and in most cases be adaptable to varying conditions encountered during the works.

2.1.3 Functional requirements

In order to begin to design any tunnel lining it is important to know and understand the functional requirements that the lining needs to achieve. There can be a wide variety of requirements, which are influenced by many factors. A tunnel lining is fundamental to most underground construction projects, usually to enable the underground space to be used as required. Of paramount importance to this is the long-term integrity of the tunnel structure, which is totally dependent on the serviceability of the lining. The major requirements of the lining may be summarised as follows.

- **Operational** Usually determined by the owner/operator and dependent on the purpose of the tunnel and how it is to be operated.
- **Serviceability** Includes the anticipated design life and the owner/ operator's view on initial capital cost versus both longer-term maintenance and shorter-term issues such as fire resistance.
- **Environmental** Including external influences from the surrounding environment, such as leakage, chemical and temperature effects, as well as the effects of the constructed tunnel on the surrounding environment, such as those from noise, vibration, changes in the groundwater regime, settlement and appearance.

2.1.4 Other factors

Risk factors will also influence the determination of the form and detail of a tunnel lining. The commercial framework under which the tunnel is to be constructed can influence the level of risk the owner, designer and contractor are willing to accept and this in turn may influence the method of construction. Risk and the way it is shared may also be relevant, particularly where new technology is involved. These factors will all play a part in defining the project requirements under which the tunnel lining will be designed, and will influence many of the technical decisions that have to be made throughout the design process.

2.2 Operational requirements

2.2.1 Tunnel function

The principal functions for which tunnels are required fall into the following categories:

Tunnel lining design guide. Thomas Telford, London, 2004

- **mining:** not covered by these guidelines
- **military:** not covered by these guidelines
- **transportation:** road
  rail
  pedestrian
  canal
- **utilities:** water supply
  sewerage
  irrigation
  cables and piped services
  hydroelectric power
  cooling water
- **storage and plant:** power stations
  liquid storage (water, oil)
  gas storage
  waste storage (e.g. radioactive)
- **protection:** civil defence shelters.

2.2.2 Function of the tunnel lining

The absolute requirements are to support the surrounding ground for the design life of the structure and/or to control groundwater inflow, without restricting the day-to-day use of the tunnel. This requirement for ground support includes the preservation of tunnel integrity under seismic conditions. Virtually all tunnels are used either for transportation (e.g. railway, road, pedestrian, water etc.) or for containment (e.g. liquid, gas or waste storage). Many will have multiple purposes and these must be determined at the start of the project in order to confirm the minimum special constraints for the design.

Figure 2.1 sets out some of the significant spatial and loading constraints that need to be considered for linings for rail, road and utility tunnels.

2.2.2.1 Access An initial assessment of access requirements will be necessary and should include evaluation of any constraints these will impose on operation of the tunnel. Typical examples of such constraints are as follows.

- *Maximum operating speed* while personnel are in road or railway tunnels.
- *Maximum flow rate* in sewers for man access.
- *Need for and effect of pumping out* water transfer tunnels and siphons for inspection and clean out.
- *Minimum special arrangements for man access and maintenance equipment.*

2.2.2.2 External loading Loadings under which a tunnel lining will be required to operate will depend largely on the tunnel use. Primary external ground and groundwater loads such as surcharge from buildings, foundations, piles and adjacent tunnels may need to be considered as well as possible accidental load cases from possible explosions or earthquakes and other seismic disturbances. Even reduction of loading in the long term from dredging operations or the like may need to be considered.

2.2.2.3 Internal loading Internal loads will also need careful consideration and these can be either permanent or transient. Some of these are likely to be relatively small by comparison with external

	Rail	Road	Utilities
Space	Structure gauge/kinetic envelope Track and track-bed Access/escape walkway Ventilation equipment, services and M & E Electrical clearances Aerodynamics (cross-sectional area and roughness)	Swept envelope Roadway Escape route Ventilation services and M & E Aerodynamics	Capacity (volume, section, grade) Hydraulic shape Access (benching/walkway) Roughness

Loading	Internal fixing for
	  Rails
	  Walkway
	  M & E
	  Signals
	  Telecommunications
	  Overhead Catenary System
	External – ground
	    – water
	Adjacent tunnels
	Shape tolerances
	Temporary construction loads
	External influences from property development (surcharge loading, piles and foundations)
	Reduction from dredging of water crossings
	Long-term creep
		Internal fixings
		  Roadway
		  Walkway
		  M & E
		  Traffic controls
		  Secondary cladding and ducting
		  Signage
		External – ground
		    – water
		Adjacent tunnels
		Shape tolerances
		Temporary construction loads
		External influences from property development (surcharge loading, piles and foundations)
		Reduction from dredging of water crossings
		Long-term creep
			Internal pressures (surcharge)
			
			External – ground
			    – water
			Adjacent tunnels
			Shape tolerances
			Temporary construction loads
			External influences from property development (surcharge loading, piles and foundations)
			Reduction from dredging of water crossings
			Long-term creep

Fig. 2.1 Major spatial and loading considerations for tunnel lining design for rail, road and utility tunnels

Tunnel lining design guide. Thomas Telford, London, 2004

loading but may need early consideration. Accidental load cases may also need to be considered such as explosions as well as temporary construction loads from possible internal compressed air. Means of fixing, for example, need to be considered, as many tunnel owners do not allow post drilling of tunnel linings. Internal pressures in water storage and transfer tunnels need to be particularly assessed as they are likely to have a major influence on the variance of loading in the lining as well as influencing the detailing of watertightness both internally and externally.

2.2.3 Availability

Assessments for Reliability, Availability and Maintainability of systems will be needed to satisfy operators that the proposed tunnel lining will perform the required functions throughout its design life, and without unplanned special intervention to correct problems. Unavoidable difficulties in accessing some tunnels when they are in use may place a 'zero maintenance' requirement during its design life on the design of the tunnel lining.

2.2.4 Hazards

Hazards will need to be identified to ensure that both personnel and the general public are not unexpectedly put in danger as a result of either construction or normal operation. Therefore, Hazardous Operations, HAZOPS, and Risk Analysis Management, RAM, studies should form an integral part of the design process (see also Section 2.6).

2.3 Serviceability and requirements

2.3.1 Durability and tunnel environment

Tunnel linings are very often difficult to access for maintenance. The external surface is always inaccessible but in most cases this surface has relatively little air contact. By comparison the internal surface may be subjected to considerable variation in:

- *temperature* (particularly near portals)
- *humidity*
- *chemicals* (such as de-icing salts).

The internal surfaces and joints therefore tend to be more prone to durability issues and due attention needs to be given to such influences as:

- *water*
- *chemical content of groundwater from seepage, effluent, road drainage* etc.

The effects of chemicals in the groundwater as well as the possible introduction of aerobic conditions due to high groundwater movement and the effects of altering watercourses need to be considered.

- *Freeze/thaw* at portals
- *Possible risk of fire* in the tunnels (see Section 2.3.3).

2.3.2 Materials

Choice of materials to be used for the tunnel lining will be influenced by the external and internal environmental conditions as well as the points detailed above. The effects of tunnelling on the external environment will be particularly important during the construction phase.

2.3.2.1 Durability In the past, the use of brick and grey cast iron has led, in most cases, to very durable linings. However, with the increasing use of reinforced concrete and ductile (SG – spheroidal graphite) iron more detailed consideration of durability is required.

2.3.2.2 Reinforcement protection The strength of concrete used in segmental linings has increased as cement and additive technology has improved. However, the use of steel bar reinforcement means that any loss of alkaline (passive) protection from the cement paste becomes much more critical and can result in early durability problems from rusting reinforcement. In tunnels where saline intrusion is present or in railway tunnels where earth current leakage or induced currents can set up electrical cells within the reinforcement, corrosion can be particularly severe. In those circumstances it may be necessary to increase concrete cover to reinforcement or consider alternatives such as higher specification linings without steel reinforcement, coated reinforcement, fibre reinforcement or, in the extreme, cathodic protection.

2.3.2.3 Corrosion protection The increasing use of steel and SG iron rather than the more traditional use of grey iron has led to a corresponding increase in the need to consider corrosion protection and 'life to first maintenance'. In these circumstances it will be necessary to consider the type of coating required to achieve long-term protection and also the materials to be used for repair, bearing in mind the generally enclosed environment and any potential toxicity and flammability of the materials.

2.3.3 Fire

Fire resistance of the lining may be a significant factor, particularly for road and rail tunnels, and this needs consideration by both the owner/operator and the designer. The requirements need to be discussed and agreed to ensure that there is a clear understanding of the potential fire load within the tunnel and how this is to be controlled. This will form part of the HAZOPS and RAM studies referred to in Section 2.2 but the consequence of these may be a need to fire-harden the tunnel lining, or at least carry out fire tests. Similarly it may be necessary to limit the incorporation and use of specific materials such as plastics or bitumens because of their potential toxicity or low flash point.

More details are given in Section 4.6 on fire resistance.

2.3.4 Design life

Many tunnel owner operators are well informed and have their own minimum requirements for tunnel linings. These take many different forms and with the growing privatisation of infrastructure ownership (in the UK) these are becoming more disparate, although the design life is typically in the range 60–150 years. Some clients have specified design life in recent years (e.g. 100 years for the London Underground Jubilee Line Extension and 120 years for the UK's Crossrail and Channel Tunnel Rail Link projects).

Practically, there are few precedents to support specifying a design life for reinforced concrete of more than the number of years in British Standard BS 8110. However, the provisions of BS 5400 are for a design life of 120 years. These design life durations may not be applicable in other circumstances. For example, a relatively short vehicle tunnel for a quarry operator may not need to be open for many years whereas a nuclear waste repository may need very long-term durability to minimise leakage. Many major sewer tunnels in London, such as the Low Level No. 1, were built in the 1860s and lined with engineering bricks, these are still in excellent condition as are the early London Underground tunnels which are over 100 years old and still perform satisfactorily.

For a more detailed consideration of design life and durability, see Chapter 4.

2.3.5 Capital cost vs maintenance

There will always be a balance to strike between the capital cost and the cost of a planned maintenance regime. The approach taken needs to be fully discussed with the owner/operator early in the design process and agreement reached as to what extent whole life costs are taken into account at the design stage.

2.4 Environmental considerations

The environment inside and outside the tunnel needs to be considered in designing a tunnel lining.

2.4.1 Internal environment

- **Water and gases** The permeability to both water and gases of the lining can have a profound effect on its operation. Leakage of groundwater can affect humidity in the tunnel and result in misting rail tunnels, increased ventilation loading or condensation on the lining and other tunnel components. Leakage into sewers will increase the cost of sewage treatment. Permeability to gases can have potentially lethal effects if poisonous or flammable gases can build up in the tunnel. Similarly oxygen deficiency in a tunnel can be fatal if proper tunnel entry procedures are not implemented and followed by maintenance personnel.
- **Materials** The influence of chemicals within the tunnel can have a considerable effect on the durability of the tunnel. For example, hydrogen sulphide from sewerage can dramatically reduce the life of concrete linings and needs careful consideration in the planning process.

2.4.2 External environment

The effects of tunnelling on the external environment will be of particular importance during the construction phase.

- **Groundwater pollution** Chemicals, particularly in grouts used for backfilling voids, water sealing and rockbolts, can pollute groundwater and watercourses. The toxicity of these materials is covered by extensive legislation.
- **Noise and vibration** Noise and vibrations can be transmitted long distances through the ground and the effects are dependent on many variables including strata, groundwater and vibrating source within the tunnel, as well as the lining type. Railway tunnels in urban environments are particularly vulnerable to noise and vibration problems and these aspects need early consideration to determine if they will affect the tunnel size or the form of the lining.

2.5 Commercial framework

2.5.1 General

The commercial framework adopted to procure a tunnel and its lining will affect or reflect the attitude of all parties involved in specifying and constructing the tunnel lining. The type of organisation procuring the project will affect the approach taken. Government agencies and statutory authorities for example are much more likely to have predirected rules or opinions on how best to procure the project. These will have been formulated through their own experiences, which may not necessarily have been for tunnels. A private organisation may take a more flexible approach. However, the commercial framework must not affect the designer's main responsibility to produce a competent, safe design, whatever the procurement method. Invariably, designers are not relieved of their responsibility by commercial and/or time pressures imposed by the client.

2.5.2 Funding and form of contract

Project funders may state specific requirements for the way in which the project is procured. Within the European Union, notices need to be placed in the Supplement to the Official Journal of the European Communities for any contract above a given financial value threshold. Project funders and/or owner/operators may require specific forms of contract which in turn will contain varying distributions of risks.

2.5.3 Method of measurement and risk apportionment

The method of measurement and reimbursement will reflect the risk apportioned between the contractor (or supplier) and the owner/client. For example, the contractor takes a more significant proportion of the risk with a lump sum or target contract. The risk apportioned may well affect the contractor's enthusiasm in suggesting and developing potential cost savings from design development or value engineering.

2.5.3.1 Lining procurement   The approach to supply of segmental linings will similarly affect the risk allocation between owner and contractor. Many linings are procured through the main contractor who then sub-contracts to a specialist lining contractor. Alternatively an owner may opt to supply linings to the main contractor by entering into a direct contract thereby accepting a greater proportion of risk. This risk may be partially offset by the appointment of a consulting engineer to carry out detailed design on the owner's behalf.

2.6 Management of risk

One of the most important underlying principles on any tunnelling project is that risk management is not optional. The responsibilities within risk management cover, but are not limited to, health and safety issues as addressed by legislation (see below). The existence of regulations and guidelines does not relieve the designer of the responsibility to design a lining competent for a particular circumstance. Although regulations must be complied with, they do not absolve designers of their overall professional responsibilities.

Designers of tunnels, and any other type of construction work, are required by legislation to consider matters of occupational health and safety. This legislation arises out of a 1992 European Directive, which was translated into UK health and safety legislation by means of the creation of:

- *Construction (Design and Management) Regulations 1994*
- *Construction (Health, Safety and Welfare) Regulations 1996.*

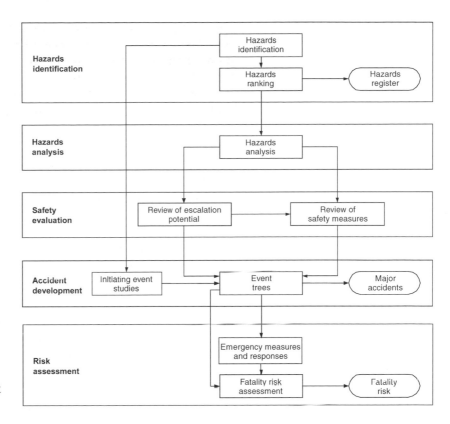

Fig. 2.2 The risk management process

To understand the extent to which occupational health and safety matters should influence design decisions and when such matters should be considered within the time-scale of the whole design process, it is necessary to look at some of the wording of the European and UK legislation. Comprehensive guidance and recommendations on occupational health and safety matters in tunnelling is contained in British Standard 6164 – 'Safety in Tunnelling in the Construction Industry'.

2.6.1 Risk Analysis and Management

Project Risk Analysis and Management (RAM) is formalisation of the common sense that most engineers employ on their projects. The difference now is the expectation, supported by CDM regulations, that key tasks be integrated into the design and construction process. The steps required to implement this are outlined in Fig. 2.2.

Risk analysis for tunnelling projects is mostly qualitative. Experience and judgement are used to identify construction methods and develop designs that meet the overall project objectives of completing a project on time, on budget and safely. Managing this process efficiently allows project managers to exercise full control at all times rather than employ reactive crisis management.

The process should focus on identifying hazards that offer a risk to the project. During design considerations these are evaluated in terms of probability and severity. Responses to the risk by the project will depend on the impact on the project objectives, safety and performance. What is important is to document how the risks either have been eliminated, mitigated or could be managed during construction. Typical residual risk documentation demonstrating this process is presented in Appendix 2 Risk Management.

Many of the RAM issues are addressed in the BTS Course 'Health and Safety in Tunnelling' and most engineers will find this informative and useful. According to Anderson (2000), implementing risk management requires two prerequisites:

- *knowledge based on education, training, competence and experience*
- *motivation arising from leadership that understands the value of risk management.*

Anderson *et al.* also goes on to summarise the key factors for an acceptable level of health and safety performance:

- *appropriate management systems*
- *practical and effective organisational systems*
- *robust engineering systems*
- *health and safety systems*
- *consideration of human factors.*

2.6.2 1992 European Directive

The Temporary or Mobile Construction Sites Directive (92/57/EEC) was drafted following a research project funded by the European Union. The research looked at the nature of accidents and ill health in the industry and the primary and other causes of why these events occur. The main findings are summarised in a number of preambles to the Directive, and two of these are relevant to the work of this document. It was found that:

> *Whereas unsatisfactory architectural and/or organisational options or poor planning of the works at the project preparation stage have played a role in more than half of the occupational accidents occurring on construction sites in the Community.*

> *Whereas it is therefore necessary to improve co-ordination between the various parties concerned at the project preparation stage and also when the work is being carried out.*

2.6.2.1 Shortcomings The first of the above clauses mentions three 'shortcomings' as adding to health and safety risks, namely:

- *unsatisfactory architectural (or engineering) options*
- *unsatisfactory organisational options*
- *poor pre-construction phase planning.*

Thus those in charge of the preconstruction planning stage should look to what can be done in terms of reducing risk by the consideration of both engineering options and organisational options.

2.6.2.2 Duties Article 2 of the Directive goes on to define three parties who are to be given health and safety duties and responsibilities, namely:

- **The Client** For whom the work is being carried out
- **The Project Supervisor** The person responsible for the design and/or execution and/or supervision of the project on behalf of the client
- **The Co-ordinator** For health and safety matters during the preparation of the project design.

2.6.3 UK Regulations of 1994

The two sets of UK Regulations mentioned above translated the Directive into what is required within the UK. Member Nations are at liberty to add matters into national law that were not part of the Directive (the UK and certain other countries did this) in addition to the minimum Directive requirements.

The Construction (Design and Management) Regulations (the CDM Regulations) should not be studied or followed in isolation, as both the Health and Safety at Work, etc. Act 1974 and the 1999 Management Regulations are also likely to be relevant.

Although the Directive did not use nor define the term 'designer' or 'design', this has been included in the CDM Regulations. The Directive's 'project supervisor' was redefined in the UK for the pre-construction stage as the 'planning supervisor', and relationships were set up between the client, the planning supervisor and the designer. They can, given certain circumstances, come from the one organisation – they do not have to be separate individuals or legal entities.

The key regulation for designers in the CDM Regulations is Regulation 13, which requires that:

> *Every designer shall ensure that any design he prepares ... includes among the design considerations adequate regard to the need to avoid foreseeable risks to the health and safety of any person carrying out construction work ...*

This regulation also contains the essence of Article 4 of the Directive, which requires the application to the design process of the 'risk hierarchy', which, in the UK, is set out in Schedule 1 of the 1999 Management Regulations. It is this framework that provides designers with the necessary template for action.

2.6.4 Joint Code of Practice for Risk Management of Tunnel Works in the United Kingdom

The British Tunnelling Society and the Association of British Insurers have jointly issued this Code of Practice in order to promote and secure best practice for the minimisation and management of risks associated with the design and construction of tunnels and associated underground structures including the renovation of existing underground structures. It sets out practice for the identification of risks and their allocation between the parties to a contract and contract insurers, and the management and control of risks through the use of Risk Assessments and Risk Registers.

2.6.4.1 Scope The scope of the code applies to the project development, design, contract procurement for tunnel construction in the UK and tunnel operation as regards any stipulated maintenance period. It also covers the impact of tunnel construction on third parties and other structures. The code excludes the operational performance of underground structures other than that included in any stipulated maintenance period.

2.6.4.2 Contents The code of practice lays down at length the requirements of the risk management process based on the commercial imperatives and details the actions and responsibilities of all parties including:

- **risk assessment**
- **risk registers**
- **client role and responsibilities:**
    appropriate technical and contract management competence
    site and ground investigations
    assessment and evaluation of project options
    project development design studies

- **construction contract procurement stage:**
  preparation of contract documentation for tendering purposes
  selection or pre-qualification of contractors for tendering purposes
  adequate time for tendering and tender assessment and evaluation
  Tender Risk Register
- **design:**
  selection and appointment of a designer
  transfer of information between designers
  the design process
  design checks
  constructibility issues
  validation of design during construction
- **construction stage:**
  pre-construction activities
  risk management procedures
  contractors' staff and organisation
  constructibility
  methods and equipment
  management systems
  monitoring
  management of change.

2.6.5 Practicalities of what designers must do in terms of strategy

Designers are charged with the consideration and elimination of foreseeable occupational health and safety risks at all stages of their design process. That is not to say that the design process has to be entirely driven by health and safety considerations, and the risks considered are limited to those that the designer could reasonably be expected to foresee. The designer is expected to have a clear view as to the likely methods of construction of his designs, and the risks and hazards associated with the various possible or likely construction processes.

2.6.5.1 Avoid risk   The first step in the risk hierarchy is, where possible, to 'avoid risk'. This requires a structured process of hazard and risk identification, and the first objective is to implement design options to eliminate hazards. Where hazards and therefore risks remain, then these residual risks have to be controlled or reduced. The end objective is to develop designs which aim to protect all those exposed to the residual risks involved in the construction work, as opposed to, for example, relying on risk control measures that only give some protection to individuals.

2.6.5.2 Provide information   There is a further important legal requirement that designers provide adequate information about any health and safety aspect of the project, structure or material that might affect the health and safety of any person carrying out work at the construction stage. This information should be passed over to those in charge of the construction works in sufficient time for them to take account of the information before construction phase, safe systems of work are devised and implemented.

2.6.5.3 Keep records   There is no formal requirement for designers to keep records of their considerations in these matters, but it would be sensible to do so as part of formal office/design

Tunnel lining design guide. Thomas Telford, London, 2004

systems in case some investigation is mounted at a later date. The enforcement authority, the Health and Safety Executive (HSE) has the power at any time to choose to audit the designer's approach to the fulfilling of his or her legal obligations. Carrying out some sort of 'risk assessment' after the design work has actually been completed and after the main engineering decisions have been made, will not satisfy the spirit of the legislation nor would such a superficial approach commend itself to the HSE.

2.6.5.4 Challenge The legislation calls on designers to 'challenge' the, perhaps accepted ways of doing things, and, where appropriate, use new approaches, new technology and fresh initiatives thereby seeking to eliminate and reduce risk and thus accidents and cases of ill health.

2.7 References

Anderson, J. and Lance, G. A. (1997). The necessity of a risk-based approach to the planning, design and construction of NATM tunnels in urban situations. *Proc. Tunnelling '97 Conference, London*. Institution of Mining and Metallurgy, London, pp. 331–340.

Anderson, J., Lance, G. A. and Rawlings, C. G. (1998a). Pre-construction assessment strategy of significant engineering risks in tunnelling. *Proc. Int. Cong. on Underground Construction, Stockholm*, June. A A Balkema, Rotterdam, pp. 191–198.

Anderson, J., Isaksson, T. and Reilly, J. J. (1998b). Tunnel project procurement and processes – fundamental issues of concern. *Proc. Conf. 'Reducing risk in tunnel design and construction', Basel*, Dec. pp. 1–13.

Anderson, J., Isaksson, T. and Reilly, J. J. (1999). Risk mitigation for tunnel projects – a structured approach. *Proc. ITA World Tunnel Congress, Oslo*, June. A A Balkema, Rotterdam, pp. 703–711.

Anderson, J., Isaksson, T. and Reilly, J. J. (2000). Tunnel project procurement and processes – fundamental issues of concern. *Tunnel Management International* **2**, Issue 7, May, 19–26.

British Standards Institution (1994). *Code of Practice for Design of Steel, Concrete and Composite Bridges*. BSI, London. BS 5400.

British Standards Institution (1997). *Code of Practice for Design of Concrete Structures*. BSI, London, BS 8110.

British Standards Institution (2001). *Code of Practice for Safety in Tunelling in the Construction Industry*. BSI, London, BS 6164.

Construction Industry Information and Research Association (1997). *Experiences of CDM*. CIRIA, London, R171.

Construction Industry Information and Research Association (1998a). *CDM Regulations – Practical Guidance for Clients and Clients' Agents*. CIRIA, London, R172.

Construction Industry Information and Research Association (1998b). *CDM Regulations – Practical Guidance for Planning Supervisors*. CIRIA, London, R173.

Health and Safety Executive (1994). *Managing Construction for Health and Safety: Construction (Design and Management Regulations)*. HSE Books, Norwich, AcoP. L54.

Health and Safety Executive (1996). *Heathrow Investigation Team. Safety of New Austrian Tunnelling Method Tunnels – with Particular Reference to London Clay*. HSE Books, Norwich.

HMSO Government (2000). *The Construction (Design and Management) (Amendments) Regulations 2000*, SI 2000/2380, London.

Ove Arup and Partners (1997). *CDM Regulations – Work Sector Guidance for Designers*. CIRIA, London, R166.

Powergen (1997). *The CDM Regulations – a Design Risk Assessment Manual*. Blackwell Science, Oxford (CD-ROM).

3 Geotechnical characterisation

3.1 General

The nature and engineering behaviour of the ground to be tunnelled through are fundamental considerations in the design of tunnel linings. In this chapter the process of geotechnical investigation and interpretation of the findings in relation to the design of tunnel linings are described. In discussing methods of ground investigation, guidance is given for the techniques and tests most appropriate to the design of tunnel linings. The particular importance of establishing the groundwater regime is emphasised.

The concept of ground appreciation, the link between the ground investigation and the tunnel lining design process itself, is introduced. A clear distinction between 'soft' and 'hard' ground is made, because this is fundamental to the design approach. As only a tiny proportion of ground is actually investigated, particular emphasis is given to looking beyond the findings of the investigation alone, to consider other features and anomalies which could exist. In other words the need to attempt to 'foresee the unforeseeable', or at least 'predict the unlikely' through risk assessment, is stressed.

The notion of the 'ground model' is useful (Muir Wood, 2000, pp. 180–182) in helping to guide the objectives of the site investigation. Its value for construction as well as design is also emphasised (p. 101). It is essential that designers control the general strategy as they know, or should know, what is relevant (p. 58). Expectations from each borehole should be understood so that any departure (the 'rogue data') then leads to instant concern, which may, in turn, affect future investigation, planning or design.

The geotechnical parameters normally required for tunnel lining design are identified in this chapter. Geotechnical test results may exhibit considerable scatter, since they are recorded on (mostly) natural geological materials, in contrast to the usually more predictable test values that would be expected from man-made engineering materials such as steel or concrete. The effects of fissures on the undrained shear strengths of clays and the influence that joints have in reducing the overall strength of rock masses are examples of the numerous factors of which account must be taken. Accordingly range and certainty of adopted geotechnical design parameter values become important issues. Principles of selection are therefore discussed in some detail, and guidance is given together with some examples of their implementation.

Over many years it has become accepted practice to provide reference ground conditions for prospective contractors of tunnelling projects at tender stage. This chapter concludes (see Section 3.8) with some advice on this topic, including distinguishing between the different circumstances surrounding designs led by the client, consulting engineer or contractor.

3.2 Ground investigation

3.2.1 Ground investigation process

British Standard BS 5930: 1999 (British Standards Institution, 1999), entitled *Code of Practice for Site Investigations* sets out the principles and much detail for conducting ground investigations for civil engineering and building schemes generally.

Tunnel lining design guide. Thomas Telford, London, 2004

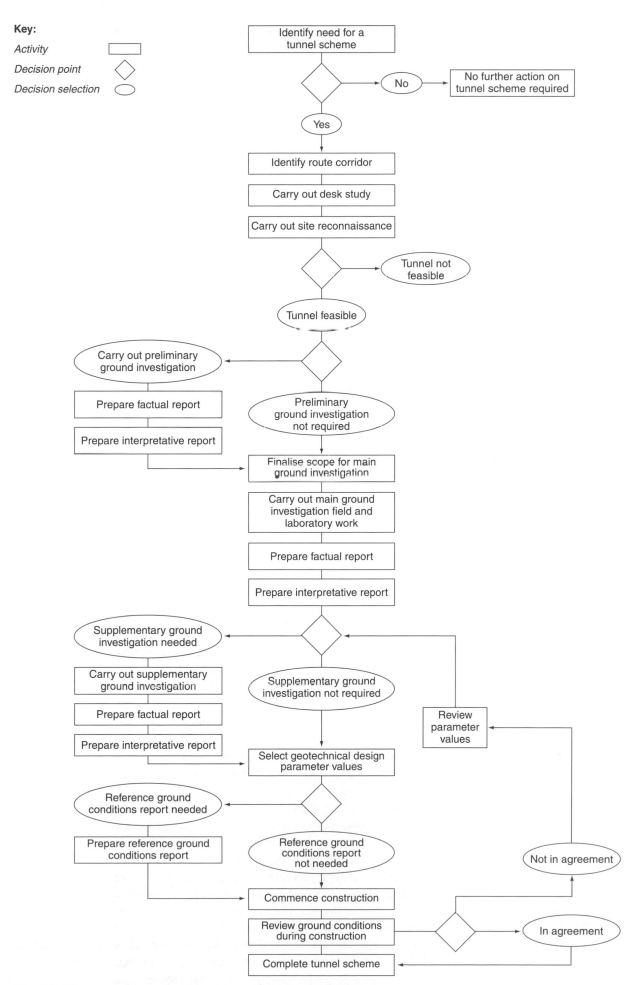

Key:

Activity ▭

Decision point ◇

Decision selection ⬭

Identify need for a tunnel scheme

No → No further action on tunnel scheme required

Yes

Identify route corridor

Carry out desk study

Carry out site reconnaissance

Tunnel not feasible

Tunnel feasible

Carry out preliminary ground investigation

Prepare factual report

Prepare interpretative report

Preliminary ground investigation not required

Finalise scope for main ground investigation

Carry out main ground investigation field and laboratory work

Prepare factual report

Prepare interpretative report

Supplementary ground investigation needed

Carry out supplementary ground investigation

Prepare factual report

Prepare interpretative report

Supplementary ground investigation not required

Review parameter values

Select geotechnical design parameter values

Reference ground conditions report needed

Prepare reference ground conditions report

Reference ground conditions report not needed

Not in agreement

Commence construction

Review ground conditions during construction

In agreement

Complete tunnel scheme

Fig. 3.1 The ground investigation process for tunnel schemes

The stages and methods of acquisition of geotechnical information for tunnel schemes can be summarised as:

- *route corridor identification*
- *desk study*
- *site reconnaissance*
- *preliminary ground investigation (if required)*
- *main ground investigation field work*
- *main ground investigation laboratory work*
- *interpretation of main ground investigation findings*
- *supplementary ground investigation (if required)*
- *interpretation of supplementary ground investigation findings (if required)*
- *selection of geotechnical design parameter values*
- *preparation of Reference Ground Conditions Report*
- *review during construction.*

The complete process is illustrated as a flow chart (see Fig. 3.1), which highlights the need to make several fundamental decisions during the process.

The ground investigation process is a specialist activity and should be directed by a suitably qualified geotechnical engineer or engineering geologist with considerable experience of tunnel schemes, under the general direction of the tunnel designer.

3.2.2 Desk study and site reconnaissance

Much information about the prevailing topography, geology, geomorphology, hydrology, hydrogeology and historical activity will already exist for most sites and tunnel routes. A desk study of the available literature, maps, aerial photographs, utility plans, existing ground investigations etc. should be carried out for all schemes. Perry and West (1996) give detailed advice on sources of existing information.

Although recent technological advances have led, in the UK, to the availability of Ordnance Survey maps which are updated frequently, there is no substitute for conducting a site reconnaissance of the route to get the best feel for the route and potential shaft and work-site locations. Normally this activity takes the form of a walk-over survey, which also provides the opportunity to examine exposures of the ground and to confirm, or sometimes question, the reliability of the desk study data. Of particular value for rural sites can be a discussion with Field Geologists of the Geological Society, to obtain a view of their notebooks or field mappings of outcrops.

3.2.3 Field investigation and testing methods

There is a large range of possible field investigation and testing methods, but many are of very limited applicability to tunnelling projects. The accompanying Fig. 3.2 lists the more common ground investigation methods and defines the appropriateness of their application for the geotechnical characterisation of tunnel projects. Further information on the methods of ground investigation themselves and advice on the frequency of sampling and testing can be found in BS 5930: 1999 (British Standards Institution, 1999) and Clayton *et al.* (1982). Figure 3.2 distinguishes between techniques that are considered fully appropriate for most tunnel projects, and those that are of limited or supplementary use.

Fig. 3.2 Ground investigation techniques

	Ground investigation		Logging or in situ testing techniques				Depth range			Tunnel components					
	Soft ground	Hard ground	Geophysical borehole logging	Logging of drilling parameters	SPT/CPT	Packer testing	Shallow	Intermediate	Deep	Portal site	Site compound	Shaft site	Shallow tunnel	Intermediate tunnel	Deep tunnel
Percussive (physical – invasive)															
Dynamic probing	■						■	▨		■	■	■	■		
Window sampling	■				■		■	▨		■	■	■	■	▨	
Static cone penetration test	■						■	▨		■	■	■	■	▨	
Percussion (shell and auger) boring	■	▨	▨		■		■	■		■	■	■	■	▨	
Rotary drilling (physical – invasive)															
Rotary percussion		■	▨	▨			■			■		■	■	■	■
Rotary open hole		■	▨	▨			■			■		■	■	■	■
Vertical rotary coring		■	▨	▨	▨	▨	■	■	■	■	■	■	■	■	■
Oriented core drilling		■	▨	▨		■		■	■	■		■	■	■	■
Excavations (physical – invasive)															
Trial pit	■	■					■	▨		■	■				
Trial shaft		■					■	▨	▨	▨		■	■	▨	▨
Trial adit							■	▨	▨	▨				▨	▨
Geophysical (non-invasive)															
Ground probing radar	▨	▨					■	▨		■	▨	▨	■	▨	
Electrical resistivity	▨	▨					■	▨		■	■	▨	■	▨	
Seismic refraction	▨	▨					■	▨		■		▨	■	▨	
Seismic reflection		▨					▨	■	■			▨	▨	■	■
Cross borehole seismic		▨												▨	▨

Key: ▨ Applicable to some ground types or project circumstances
■ Fully applicable

3.2.4 Laboratory testing methods

The range of possible laboratory tests is considerable, but those tests directly applicable to the design of tunnelling projects are more limited. Tables 10 and 11 of BS 5930: 1999 list common laboratory tests for soil and rock respectively. Figure 3.3 lists

Type of test	Parameters obtained	Symbol	Normal application					Tunnel applicability
			Soft ground		Hard ground			
			Cohesive soil	Granular soil	Mixed soil	Weak rock	Strong rock	
Bulk density	Unit weight	$\gamma\gamma'$	✓		✓	✓	✓	Overburden pressure
Maximum and minimum density	Maximum and minimum density	$\gamma_{max}, \gamma_{min}$		✓				Overburden pressure of ground or soil
Moisture content	Moisture content	w	✓	✓	✓	✓		Type and state of ground
Specific gravity	Specific gravity	G_s	✓	✓	✓	✓	✓	Overburden pressure
Plasticity	Liquid and plastic limits, plasticity and liquidity indices	LL, PL, PI, LI	✓		✓			Type and state of ground
Particle size distribution	Proportions of soil composition		✓	✓	✓			Type of ground
Unconfined compression	Unconfined compressive strength	q_u				✓	✓	Rock strength
Point load index strength	Point load index strength, unconfined compressive strength	I_p, q_u				✓	✓	Rock strength
Undrained/drained triaxial compression	Undrained shear strength	C_u, S_u	✓		✓	✓		Soil/rock strength
Consolidated undrained/drained triaxial compression	Effective stress shear strength (not with an undrained test)	c', ϕ'			✓			Soil strength
Shear box	Angle of shearing resistance	ϕ, ϕ'		✓	✓	✓	✓	Frictional strength of soil grains and rock joints
Odometer/one-dimensional consolidation	Coefficient of compressibility/ vertical drained deformation modulus	m_v, E_v'	✓					Soil stiffness
Laboratory permeability	Coefficient of permeability	k		✓	✓			Soil permeability
Poisson's ratio	Poisson's ratio	ν	✓			✓	✓	Ground stiffness
Chemical analyses	pH Sulphate content Chloride content	pH SO_3 Cl	✓	✓	✓	✓		Lining durability
	Chemical contamination of soil, rock or groundwater		✓	✓	✓	✓	✓	Lining durability
Abrasion							✓	Excavatability
Shale durability						✓	✓	Softening susceptibility

Fig. 3.3 Laboratory tests to obtain geotechnical parameters for tunnelling schemes

Tunnel lining design guide. Thomas Telford, London, 2004

tests that should normally be considered for tunnel schemes, together with the parameters obtained and the extent of their application to design in soils and rocks.

3.2.5 Factors to consider in selecting investigation methods and scope

While there are clearly a number of field and laboratory investigation and testing methods of direct relevance to tunnel schemes, the methods and scope of investigation applicable to any particular scheme will be governed by many project specific factors. This is a large subject in its own right and it is beyond the scope of this Guide to go beyond an introduction to the principles of the decision making process and a list of the main factors.

British Standard BS 5930: 1999 provides general guidance on the location, spacing and depth of explorations. Glossop (1968) has expressed a valuable principle:

> *If you do not know what you should be looking for in a site investigation, you are not likely to find much of value.*

Allied to this is the principle of a strategy for site investigation defining a purpose for every single exploratory hole, soil sample or rock core, field test and laboratory test undertaken. This should ensure that optimum value for money has been obtained from sound, engineering-based considerations. During the time that information is obtained the strategy may change and the programme will be varied appropriately.

A phased geotechnical investigation programme is of great benefit in following these principles. The findings of a preliminary investigation enable the main investigation to be better tailored to suit the expected ground and groundwater conditions, while a supplementary investigation can be planned to sweep up remaining doubts and knowledge gaps left after assessing the results of the main investigation. On the other hand, early boreholes may, with economy, serve the purposes of a later phase as, for example, the installation of piezometers.

In recognising that each combination of project and site is unique, there is no set of rules that can be applied rigidly in every case, but the following aspects must be considered:

- *character and variability of the ground*
- *nature of the project*
- *need for and scope of a preliminary investigation*
- *location of exploratory holes*
- *spacing of exploratory holes*
- *depth of exploration*
- *potential for ground contamination.*

Schemes involving tunnels and shafts have particular considerations. Boreholes should be located off the tunnel line so as not to interfere with subsequent construction and they should always be sealed through aquicludes. Also, it is important to take boreholes to an adequate depth below the proposed invert level, both because subsequent design changes could cause a lowering of alignment, and because the zone of influence of the tunnel may be extended by the nature of the ground at a greater depth.

A useful discussion on this subject is Section 4.3 of *Tunnelling: Management by Design* under the heading of 'How much site investigation?' (Muir Wood, 2000).

3.3 Soil and rock description and classification

3.3.1 Soil

Soil descriptions are derived from field and laboratory tests and observations of disturbed and undisturbed samples. British Standard BS 5930: 1999 gives full details of the approach to be followed in identifying and describing soils. Information relating to the identification of the main soil types is listed in Fig. 3.4, but it should be noted that soils are almost invariably composed of combinations of different particle sizes. The description system in Fig. 3.4 is principally applicable for earthworks purposes, and may not be wholly appropriate for descriptions to be evaluated in the design of tunnel linings. For example, the description of clay would normally include a statement of its consistency, being an indication of its intact shear strength. But, the description itself would not usually give any indication as to the degree of over-consolidation of the clay, nor, consequently, the coefficient of earth pressure at rest.

Soil type	Soil name	Particle size: mm
Coarse	Boulders	>200
	Cobbles	200–60
	Gravel	60–2
	Sand	2–0.06
Fine	Silt	0.06–0.002
	Clay	<0.002

Fig. 3.4 General composition of ground types

3.3.2 Rock

For a rock mass, classification systems (e.g. BS 5930, ISRM, GSL and IAEG) provide a statistical means of describing its characteristics and of predicting how it will respond to excavation. They are typically used, as an initial step, to select the support required to maintain short- or long-term stability.

In their simplest form, classification systems relate the type, quantity and dimensions of the support actually installed in a number of existing projects to the dimensions of these projects, without attempting to produce a general numerical relationship from this data (Carlsson *et al.*, 1992, for example). Other systems go beyond a simple graphical comparison of real data, to produce a numerical definition of the rock mass. By far the most internationally recognised of these are the Rock Mass Rating (RMR) (Bieniaski, 1976) and the Tunnel Quality Index systems (Q) (Barton *et al.*, 1974; Grimstad and Barton, 1993). Both of these adopt a number of parameters that can be either measured or visually estimated from borehole cores. The selected parameters are each given a numerical rating and these ratings are combined to produce a single number, considered to represent the mass characteristics of the rock through which tunnelling is to take place. The amount of support required can be determined from a series of formulae, or else directly from a graphical menu. Within the Tunnel Quality Index (Q), the Stress Reduction Factor (SRF) has been improved on the basis of 1000 recent case studies.

Classification systems can also be employed to provide the rock mass strength and stiffness parameters needed for a numerical analysis. The rock mass can be analysed with the assumption that it is either continuous or discontinuous. If continuous, the visually estimated or laboratory determined material parameters must be reduced to represent the strength and stiffness of the rock mass as a whole. If discontinuous, parameters are required to model both the discontinuities and the blocks of continuous material between the discontinuities.

Some continuum parameters, such as the rock mass modulus, can be derived from both the RMR and Q systems. The Geological Strength Index (GSI) (Hoek *et al.*, 1995, Hoek, 2000) can also be used to provide the full range of parameters needed for both the Hoek–Brown and Mohr–Coulomb failure criteria. The parameters required for discontinuum analyses can be derived from various papers produced by Barton and others (Barton *et al.*, 1992; Barton and Choubey, 1977; Barton and Bandis, 1982, for example). In most cases linings of rock tunnels are designed assuming that the ground is a continuum and that the movement of the ground towards the excavation will load the lining.

Most tunnels for civil engineering projects are excavated at moderate depths. In hard rock it can be assumed that the strength of the rock will be very much greater than the ground stresses imposed upon it. In this case discontinuities will control the stability of the rock mass, and a support system that increases the strength and stiffness of the discontinuities is appropriate. However, should the rock mass strength reduce to a point at which the walls of the excavation begin to displace, the support pressure provided by a continuous structural lining is likely to be required.

Accordingly it is recommended that direct selection systems, such as RMR and Q, be used only when the strength of the rock mass adequately exceeds the ground stresses. Otherwise the rock mass should be treated as a continuum. Some rock types, for example slates, are markedly anisotropic and can have much higher ground stresses in one direction than in another. Determination of the in situ stresses in three directions becomes fundamentally important in such cases.

3.4 Groundwater identification in soils and rocks

The groundwater regime should be explored from the outset of any tunnel investigation. During the desk study stage of the process all available well records, hydrogeological maps and any other relevant information should be researched as a matter of course.

Because of the practical difficulties of measuring accurately the groundwater levels and pressures from investigation boreholes themselves, it is normal to install standpipes and/or piezometers to record levels on equilibration following the investigation. These devices also permit groundwater levels to be monitored for some considerable time afterwards, so that longer term changes to seasonal or weather-related effects can be measured.

Groundwater effects differ for soft ground (soils) and hard ground (rocks). In soils and jointed rocks the principle of effective stress applies, so that groundwater pressures are directly mechanically related to total and effective soils stresses. The cemented, indurated nature of rocks means that effective stress principles do not apply to the constituent rock grains. However, in rocks, groundwater will infiltrate and lubricate joints (depending on joint filling), fissures and other discontinuities with resulting build up of groundwater pressure and potentially heavy flows into excavations.

Some types of ground are sensitive to the effects of contact with free water upon excavation and may be considerably weakened. Examples include fine-grained soils and rocks (clays, shales, mudstones), which are or can readily become slurry, silts and fine sands, which can exhibit 'quick' conditions, and certain salt formations which may dissolve. There is also the possibility of 'piping' as a result of a local increase in pressure gradient.

The chemical and physical nature of the groundwater can also have an effect on the design of tunnel linings. Groundwater chemistry can affect the durability of linings. In exceptional circumstances account may need to be taken of groundwater temperatures.

A fully defined groundwater regime including the predicted levels of the phreatic surface (the upper surface of the groundwater table), groundwater pressure-depth profiles (sometimes hydrostatic, sometimes non-hydrostatic) and expectations of perched groundwater tables and artesian situations, should be developed as an integral part of the complete geotechnical model of the ground throughout the tunnel route.

During construction, observations of the actual groundwater conditions should be made to check the validity of predictions where these are critical to the tunnel lining design. The Site Quality Plan referred to in Chapter 9 of this Guide should include a process to ensure that these observations are made. Such observations will include monitoring the flow of groundwater into the underground excavations, noting flows from any temporary groundwater drawdown operations around the excavated zone, and recording flows from probe holes drilled ahead of the tunnel face. If there are significant critical differences between previously assumed and actual conditions, the lining design must be reviewed and, if necessary, revised. It may be too late to do so, in which case additional ground treatment is most likely to be required.

3.5 Ground appreciation – link between investigation and design

3.5.1 Interpretation process

Ground appreciation, in other words the interpretation of ground investigation data, forms the essential link between the factual information derived from desk studies, field and laboratory investigations, and the commencement of the tunnel lining design process itself. British Standard BS 5930: 1999 (British Standards Institution, 1999) sets out, in general terms, the contents of a Geotechnical Factual Report (GFR).

Interpretation involves the definition, description and quantification of the ground in a form that is relevant and readily available for the tunnel lining design team. This information is normally provided in the form of a Geotechnical Interpretative Report (GIR). This can be prepared in full once the GFR, which contains all the findings of the field and laboratory work, has been completed. Typically the GIR would contain the following sections:

- *executive summary*
- *introduction*
- *outline of the proposed scheme*
- *definition of route corridor*
- *findings of desk study and site reconnaissance*
- *identification of route and alignment options*
- *summary of the ground investigation work*
- *description of ground and groundwater conditions*
- *interpretation of ground conditions in relation to the design and construction of the proposed scheme*
- *recommendations for design and construction and further ground investigations*
- *conclusions.*

A GIR may, and should, be prepared in conjunction with every phase of ground investigation.

The title Geotechnical Design Summary Report (GDSR) has been used on occasions, for a report containing the required geotechnical design parameters. However, this Guide does not recognise a GDSR as a report separate from the GIR; the GIR would normally be expected to include this information.

The GFR and GIR will form the basis for the Geotechnical Baseline Report (GBR) where appropriate, as discussed in Section 3.8. However, the GBR serves a different purpose and should be an entirely separate document. This is because the GBR provides the reference ground conditions for the tunnel contract, and should be provided to all parties involved in the scheme. In contrast, the GIR is prepared for the primary benefit of the design team.

3.5.2 Soft ground, hard ground and transition

In actuality there is a continuous transition from the softest, loosest or most unstable of subsoil through stiff or dense but unconsolidated ground into weak, highly fractured or weathered rock and then to strong, massive rock. However, it is traditional tunnelling practice to think in terms of tunnel schemes being in 'soft ground' or 'hard ground', regardless of this transition. Fundamental decisions, for example whether a sprayed concrete lining may be used, or which type of tunnel boring machine could be employed, are made by tunnel scheme designers and contractors based on a simple distinction between 'soft' and 'hard' ground, but often the ground conditions are more complicated and the distinction is rarely that simple.

Section 1.4 defines these terms for the purposes of tunnel lining design. Section 3.3 describes the process of identifying and classifying ground, but this does not give tunnel engineers immediate and clear definitions of 'soft' and 'hard' ground.

Broadly speaking, all types of soil, and weak rocks, would normally fall into the category of 'soft ground'. The weak rocks include the poorer grades of chalk, weak mudstones, and weakly cemented and/or highly fractured sandstone. 'Hard ground' would generally comprise all other forms of rock.

3.5.3 Groundwater behaviour

Careful evaluation of the presence and pressures of groundwater, and their potential influence, is of utmost importance in the design of tunnel linings. Unforeseen groundwater phenomena frequently cause underground construction problems, which result in lengthy delays and excessive increases in cost, and may require significant, costly alterations to ground support. Even small groundwater leakage into tunnels can, in the long term, cause structural damage as well as lowering groundwater tables with possible effects on third parties.

The three principal situations where differing groundwater circumstances can have a profound effect are as follows.

- **Encountered** Groundwater is encountered where none is expected.
- **Anticipated but effect unknown** The presence of groundwater is anticipated but its effect on the behaviour of the ground and the planned construction method is not adequately forecast.

- **Anticipated but quantity unknown**   Groundwater is anticipated but the actual quantity of inflow greatly exceeds expectations.

The presence and nature of the groundwater regime impacts upon the tunnel lining design in the following ways.

- **Range of pressures**   Establishing the range of groundwater pressure design cases to be assessed, including at construction stage as well as long term.
- **Waterproofing**   Establishing the requirements for tunnel waterproofing. The decisions as to whether or not to incorporate gaskets, membranes, etc. are fundamental and, in addition to consideration of tunnel lining design, will also depend on consideration of the operational needs of the tunnel.
- **Flotation**   Ensuring that the risk of flotation is avoided.
- **Effects on third parties**   Due to inflow to the tunnel or outflow from it such as deterioration of foundations, well drying or pollution.
- **Effects on ground of free water**   Establishing whether the properties of the excavated ground could change upon exposure to free water, for example softening, swelling (e.g. gypsum), developing 'quick' conditions, dissolving or leaching.
- **Drainage**   Providing an appropriate operational tunnel drainage system that deals with long-term groundwater inflows if expected.

It is of paramount importance not to overlook groundwater behaviour and control arrangements during the design and construction stage risk assessment processes.

3.5.4 Foreseeing the unforeseeable

It is not uncommon for a geotechnical investigation, even one based on the most rigorous and comprehensive of programmes, to fail to reveal ground conditions that depart markedly from those indicated for the tunnel route and based on the exploratory holes and tests actually conducted. Nevertheless the interpretative report must endeavour to consider, as thoroughly as possible, a picture of ground conditions as a whole.

It is a crucial function of the directing geotechnical specialist to endeavour to identify any potential anomalies that the investigation's findings have not detected. This exercise is termed 'foreseeing the unforeseeable'. Basically it entails considering the types of geological and man-made inclusions that could possibly be present between the locations of exploratory holes and tests. Examples of such inclusions are:

- *scour hollows in sedimentary cohesive strata, infilled with subsoil of a different origin*
- *groundwater bearing lenses or layers of granular material within cohesive strata*
- *cavities in rocks, for example swallow-holes in chalk, solution features in limestone, sometimes partly or fully infilled with gravels, sands, silts or clays*
- *man-made obstructions, including piles, deep drains or earlier tunnels*
- *infilled excavations – man-made or natural surface hollows*
- *chemically contaminated ground*
- *old mine workings*
- *methane or other hazardous gases*

- *underground, or 'lost', rivers or artesian/sub-artesian groundwater ingresses*
- *changes between hard and soft ground conditions that have major effects on tunnel progress rates*
- *variation in rock head or competent ground cover.*

This list is by no means exhaustive, nor would all of these types of feature be reasonably anticipated for every tunnel route. Rather, it is a matter of standing back from the geotechnical investigation data, reviewing carefully the desk study and site reconnaissance findings, and then applying common sense to anticipated further potential surprises that the ground could offer.

Although the effects of earthquakes or earth tremors are generally limited on deep, or even most shallow, tunnels, it should be recognised that these should be considered in areas subjected to ground movements at mines or nearby natural earthquakes.

3.6 Geotechnical parameters required for tunnel lining design

3.6.1 Geotechnical design parameters and their application

The required geotechnical design parameters should be defined in the section of the GIR dealing with the interpretation of ground conditions in relation to design and construction (see Section 3.5.1).

For tunnel schemes these parameters are mostly obtained from the information contained in the GFR, namely soil and/or rock descriptions and the results of in situ and laboratory tests. Figure 3.4 lists the geotechnical parameters that are normally obtained and states how they are applied to tunnel lining design. Because of the complex nature of geological materials (see Section 3.6.2), the values need to be chosen with great care. Some of the principles involved in choosing values are discussed below, and some examples of the decision-making processes involved are presented.

3.6.2 Range and certainty

Characterising the properties of man-made materials used in the structure of tunnels (most commonly these are concrete and steel) and the naturally occurring geological materials of the ground (soils and rocks) are fundamentally different processes. Man-made materials are manufactured in controlled conditions either in factories as with steel products and precast tunnel linings, or batched for in situ construction, as with ready mixed or sprayed concrete. Their required engineering properties can be specified in the design process with a high degree of confidence in their performance. In complete contrast, the engineering properties of geological materials have already been dictated by nature. It follows that there is inevitably a considerable scatter and uncertainty in these properties.

This leaves the tunnel designer with the problem of how to select the most appropriate geotechnical design parameters from an extensive array of soil and rock descriptions, and in situ and laboratory test results. There is no simple, universal solution to this problem, since every tunnel project has its own unique set of design requirements and ground conditions. However, it is possible to set out some general guiding principles and give some simple examples of the selection process.

3.6.2.1 Obtaining and utilising relevant information   The obtaining of sufficient relevant information is addressed in Section 3.2.5, while Section 3.5.4 stresses the need to look for potential surprises

Geotechnical design parameter	Symbol	Application to tunnel lining design
Soil and/or rock description from rotary coring		Defines types of ground
Grade of rock	Q, RMR	Extent of ground support
Percentage core recovery and core condition	TCR, SCR, RQD	State of weak rock or hard ground
Unit total and effective weights	γ, γ'	Overburden pressure
Relative density of coarse grained soils	D_r	State of natural compaction of cohesionless soft ground
Moisture content	w	Profiling of property changes with depth
Specific gravity	G_s	Type of ground
Plasticity and liquidity indices	LL, LP, PI, LI	Type and strength of cohesive soft ground
Particle size distribution		Composition of soft ground
Unconfined compressive strength	q_u	Intact strength of hard ground
Point load index strength of lump	I_p	Intact strength of hard ground lump
Axial and diametral point-load index strengths	I_a, I_d	Axial and diametral intact strengths
Undrained shear strength	C_u, S_u	Shear strength of soft ground
Effective stress shear strength	C'	Long-term cohesion of soft ground
Angle of shearing resistance	ϕ, ϕ'	Long-term shear strength of cohesive soft ground Short- and long-term shear strength of cohesionless soft ground
Drained deformation modulus	E'	Long-term stiffness
Poisson's ratio	ν	Influences stiffness values
Coefficients of effective earth pressure	K_o, K_a, K_p	Ratio between horizontal and vertical effective stresses at rest, active and passive
In situ stresses in rock	σ	Magnitude of principal stresses in rock in three directions
Permeability	k	Characteristic ground permeabilities and variations. Waterproofing
pH, sulphate and chloride contents	pH, SO_3, Cl	Concrete and steel durability
Chemical contamination		Extent of ground contamination
Abrasion		Rate of cutter tool wear

Fig. 3.5 Common geotechnical parameters and their applications in tunnel design

Tunnel lining design guide. Thomas Telford, London, 2004

not identified by the geotechnical investigation work. The governing principle in setting scopes of phased field and laboratory investigations should be that all exploratory holes and tests have the purpose of obtaining information of direct relevance to the design and construction of the tunnel scheme. If the question 'How is the information from this borehole or test to be employed?' cannot be readily answered, then it is not good value for money to undertake this particular activity.

For tunnel lining design it will be necessary to obtain sufficient information on geotechnical design parameters. The amount of information required will vary greatly, but Fig. 3.5 lists those parameters most often required. The degree of detail will essentially depend on the nature of the design analysis. Establishing the geotechnical properties of some geological strata can be undertaken with confidence, arising either from previous tunnelling experience within them, or from the fact that these strata are known to be relatively uniform. Arguably, London Clay is a good example of the former (previous tunnelling experience), and Mercia Mudstone of the latter (uniform strata). When expecting such strata it might be decided to have fewer boreholes, perhaps directing a proportion of the resulting cost saving into undertaking more sophisticated sampling and testing, aimed at establishing a better understanding of small strain stiffness behaviour, for instance.

Unfortunately, a high proportion of tunnel scheme sites will depart markedly from these more uniform ground conditions. The geotechnical specialist in charge must be expected to recognise such situations. The approach in these cases must be to ensure that sufficient boreholes are carried out to create a comprehensive picture of the ground conditions as a whole as well as variations along the tunnel route. Having a number of phases of ground investigation is advantageous.

All in situ and laboratory tests should serve a purpose. For example, there is no merit in undertaking just a handful of moisture content determinations on disturbed samples of different types of soils, since it would be impossible to relate the results of these tests to one another. Nor would there be enough determinations to be able to examine the scatter of results in any one particular soil type. In any event the selection of disturbed samples for these tests may well have been inappropriate, because such tests could have given unreliable results due to disturbance. In contrast, the undertaking of a large number of moisture content determinations (a relatively inexpensive form of test) on undisturbed samples in specific strata, in order to profile this parameter with depth and between borehole locations, could prove to be good value for money if capable of correlation with more significant properties.

The investigation and design processes should be interactive. The information sought for the design may become more focussed as design options are narrowed.

3.6.2.2 Identifying patterns There should normally be strong inter-relationships between the types of ground identified and the values of the test results, as well as between the different test results themselves. The geotechnical specialist directing the work should examine these patterns, and conclude as to whether they conform to expectations or not. Not only should this exercise underpin the reliability of the available geotechnical information as a whole, but also any identified inconsistencies should draw attention to anomalies and perhaps the need for further investigation and/or testing. For example, coarse-grained soils would be expected to exhibit much higher permeabilities than fine-grained soils, but there is a huge amount of scatter in permeability tests by their very nature. A full order of magnitude difference between tests in the same geological stratum (i.e. one test result showing about ten times more permeability than another) might be reasonably expected, but two orders of magnitude difference between tests could signal that the ground conditions might not have been fully identified. Further investigation might lead to the conclusion that there are, in fact, permeable sandy lenses within a predominantly clay stratum in which none were initially anticipated.

Muir Wood (2000) relates the example of how the several phases of investigation for the Channel Tunnel project evidenced cyclical variations in the clay to carbonate ratios in the chalk marl of the Lower Chalk. There were associated patterns of varying permeability. This information assisted decisions on tunnel alignment and predictions of changing ground conditions ahead of the advancing tunnel faces.

3.6.2.3 Maximum and minimum bound design parameters The pattern identification process described above should help to determine the level of confidence that can be applied to the results for each test. No single test result should be discarded as rogue and unreliable just because it is much higher or lower than all the other values. Such a result may be because the test specimen was disturbed, but it could be a genuine marked departure from the norm. For example, unconfined compressive strengths recorded on specimens of fully intact rock could be many times greater than a result obtained from a specimen of the same rock that at first sight is intact and homogeneous but, on testing, is found to contain a thin seam of much weaker material. The presence of such seams could be of great significance to the engineering behaviour of the rock mass as a whole. Any such finding should alert the geotechnical specialist to the need for further investigation or sample re-examination.

Given the diversity of geological origins and tunnel schemes, there can be no hard and fast rules for the selection of values for each geotechnical design parameter, nor for their ranges. It is common practice to present the results of in situ and laboratory tests in the form of graphical plots, and then to undertake statistical analyses to derive geotechnical design parameters. While this can be useful, it can also lead to the selection of inappropriate values for these parameters. This is because most often what is required is the selection of maximum and/or minimum parameter values. Sensitivity checks may also be beneficial. The process of sensitivity analysis is introduced in Section 3.6.2.4.

The determination of design undrained shear strength in London Clay forms a good example of how to select maximum and minimum bound values with confidence. First, having established that the test results have a common basis according to the test method used (i.e. same diameter of test specimens, same rate of shearing, etc.), the undrained shear strengths are plotted versus depth (or versus reduced level if more appropriate) for groups of boreholes deemed to show similar profiles from the pattern identification process.

Next, the precise purposes for which the parameter is required are identified. This helps define whether minimum as well as maximum values are required. For the selection of excavation plant, maximum undrained shear strength values would be needed. However, minimum design undrained shear strengths would normally be required for the design of permanent tunnel linings.

The next step, therefore, is to plot the minimum undrained shear strength lines on the depth plots. Since the design values will vary with depth, and to an extent with location depending upon the pattern identification outcome, it is to be expected that the lines will be sloping (generally, but not necessarily, strength increasing with depth). It is also likely that there will be several plots representing different parts of the tunnel route. It may also be important to plot spatial and directional variation to determine patterns.

A judgement must be made as to whether minimum lines can be drawn 'to the right of' some of the test results (i.e. to reflect design values that are higher than some of the actual results at certain depths). The geotechnical specialist must take account of the scheme proposals in making this judgement. But in this regard the sensitivity analysis process (see Section 3.6.2.4) should provide the confidence to be able to draw a minima line 'to the right of' the excessively conservative line which is 'to the left of' all the test results.

3.6.2.4 Sensitivity analysis Geotechnical design parameter sensitivity analysis is a powerful tool that the designer can use to address the 'what if' situation. The process of design parameter selection should have followed the guidelines of data processing described in the above sections, culminating in the choosing of minimum and maximum bound parameters. The tunnel lining designer should then have the parametric information needed to develop a design. If desired, the design can then be analysed to see if it is still satisfactory, albeit with a lower margin of safety, under a combination of what are termed 'worst credible' geotechnical circumstances. If the variation is on a small scale, a stability analysis is unlikely to be based on the most pessimistic values.

A simple example of this would be to examine the tunnel lining design in the context of the very lowest undrained shear strength test results (those falling 'to the left of' the minimum design line) identified in the example described in Section 3.6.2.3. In this case it is to be hoped that the selection of the appropriate lower bound design line would itself have been correct, leading to the demonstration of a satisfactory margin of safety being maintained.

A related but potentially more critical example of an area in which geotechnical design parameter sensitivity analysis has a role is in evaluating the sensitivity of a tunnel lining design to wide ranging changes in the coefficients of earth pressure (K_o – at rest, K_a – active, K_p – passive). Depending on circumstances, the actual confining pressures that apply all around circular tunnels can differ markedly from the theoretically predicted values, both in absolute terms and in the ratio between horizontal and vertical effective stresses. The effects on the tunnel lining's margin of safety, in a structural manner, can be dramatic if, for instance, there is next to no pressure confining the sides of the tunnel but there is a full overburden pressure acting on the tunnel's crown. Such extremes of confining pressure can be readily investigated by sensitivity analysis, having first determined the practical limits of variation in the coefficients of earth pressure. At times this analysis may lead to surprising, even alarming, conclusions entailing redesign, but this will be infinitely more desirable than having a lining failure following construction, which has happened on occasions.

It is also worthwhile emphasising the importance of the appropriate selection of one particular geotechnical design parameter, that of stiffness. This is evaluated as the ratio between applied stress and resulting strain, but the values adopted may be inappropriate if they relate to high strain levels, whereas many tunnels are constructed with the opportunity for the ground to develop only small strains. Again, sensitivity checks can be undertaken to understand the consequences of differing strain levels when selecting stiffness values. Since strain magnitude will vary throughout the ground mass, precise analysis will need to consider this fact under the same circumstances, but will not necessarily be adopted for the initial simple conservative method.

3.7 Ground improvement and groundwater control

3.7.1 Changes in water table
A lowering of the groundwater table would be expected to result in a general improvement in soil matrix strength, stiffness and stability, and lead to easier tunnelling conditions. The effects differ for granular and cohesive soils, as described below. Conversely, a rise in groundwater table would generally give rise to more difficult tunnelling conditions in soils. Changing groundwater levels have little effect on rock mass, but can dramatically influence the effective shear strength of rock joints.

3.7.2 Effects on ground parameters
The primary effect of lowering the groundwater table and reducing porewater pressures is to cause an increase in the effective stress between particles of soil and fragments of rock. This gives rise to higher shear strength and stiffness in cohesive soils, although their relatively low permeability means that these changes may take some time to manifest themselves. The stability of free draining granular soils and rocks is improved through the increased effective stress giving rise to higher shear resistance. In contrast, the effect of a higher groundwater table is to decrease the effective stress and consequently weaken the ground.

3.7.3 Methods of ground improvement
3.7.3.1 Permeation grouting Once the grout sets, the strength and stability of soil and rock matrices is much improved. Care must be taken not to introduce excessive amounts of grout, which could actually cause the ground mass to heave.

3.7.3.2 Compaction The placing in layers and thorough compaction of selected soil and/or rock fill produces an 'engineered' dense soil matrix, which has an advantage over natural ground in that its properties are known and can be fully relied upon. It should be noted that compaction plant has a limited depth of influence, and cannot fully compact the basal zones of layers that are more than a few hundred millimetres thick.

3.7.3.3 Jet grouting and vibro-replacement In different ways, these techniques provide both vertically strengthened columns of subsoil, and densified and strengthened soil masses between the columns.

Tunnel lining design guide. Thomas Telford, London, 2004

3.7.3.4 Dynamic compaction This technique is not normally effective in cohesive soils, but can improve the relative density (i.e. the actual soil dry density relative to the laboratory determined minimum dry density, expressed as a percentage of the difference between laboratory determined maximum and minimum dry densities) of granular soils by some 30% to 50% typically. In turn this gives a much increased shear strength.

3.7.4 Methods of groundwater control

3.7.4.1 Exclusion methods The effect on the ground is to reduce the porewater pressure to zero, so that the effective stress becomes equal to the total stress, having a direct beneficial effect on strength, stiffness and stability.

3.7.4.2 Low-pressure compressed air Reduced groundwater flows result from a reduction in the pressure difference that drives groundwater through granular soils.

3.7.4.3 Ground freezing Soils are markedly strengthened and stiffened, leading to much increased stability. With ice being about 10% less dense than water, care must be taken not to provide so much freezing that significant swelling occurs, giving rise to adverse ground heave effects.

3.7.4.4 Dewatering The mechanism giving rise to the strength, stiffness and stability benefits deriving from dewatering is described above. However, it must be borne in mind that an increase in effective stress gives rise to a consolidation of ground. A further consequence of dewatering is the progressive removal of finer particles from the soil matrix. The potential that these effects will give rise to ground settlement must always be considered.

3.8 Reference ground conditions

Section 3.5 explained that the Geotechnical Interpretative Report (GIR) is normally the document to which the design team would refer for geotechnical design parameters, for example for the detailed design of tunnel ground support. This section considers the rationale for having, for all tunnel schemes, a Geotechnical Baseline Report (GBR), in which reference ground conditions for contractual purposes are to be found. As stressed in Section 3.5, the GIR and GBR are separate documents serving different needs. The GBR draws upon the contents of the Geotechnical Factual Report (GFR) and the GIR but is likely to be a significantly different document, since it is primarily for inclusion in the contract, rather than for the needs of the designer.

The definition of reference ground conditions is important for tunnel contracts, in view of the tunnelling industry's background of overspends directly attributable to allegations of actual ground conditions differing significantly from those anticipated. Historically it has often proved difficult to quantify contractors' unforeseen ground conditions claims accurately, because of the lack of well-defined benchmark conditions agreed at contract outset between Client, Contractor and Engineer. The GBR is a logical tool for addressing this problem.

The concept of having a GBR with contractual status is not new, as it has been usual practice in the United States of America for a long time, and is becoming more common in the United Kingdom, where it has been advocated for more than 20 years (Essex, 1997) although not sufficiently carried out in practice. It is strongly advised that a GBR should be prepared for all future UK tunnel schemes. The GBR should form a part of the Contract, irrespective of the type of Contract. It would therefore be readily and openly available to all parties associated with the tunnel scheme. Examples of the use of GBR include the explicit removal of a serious risk of low probability from the responsibility of the Contractor as for the Heathrow Airport Cargo Transfer Tunnel, and its being used as the basis of 'zoning' for ground support needs and for payment related to the stated ground quality.

All factual and relevant interpretational data assembled for a project should be considered as a valuable project resource, for design and construction. The GIR should therefore be made available to the Contractor for information and, possibly, without warranty unless it provides the basis for reference conditions.

The role of site and ground investigation in the minimisation and management of risk, and the use of resultant information in correct tender documentation, is addressed in the BTS/ABI Joint Code of Practice (see Section 2.6.4).

The GBR's primary purpose is to establish a definitive statement of the anticipated geotechnical conditions ahead of tunnel construction, as a baseline for contractual reference if subsequently required. The contractual framework to be adopted for construction will reflect the Client's approach to and acceptance of risk, the consequences of which being advised by an experienced tunnelling engineer. It is normal that risks associated with conditions consistent with or less adverse than the baseline are allocated to the Contractor, and the Client accepts those risks significantly more adverse than the baseline. Essex (1997) discusses the subject of risk allocation in detail.

In circumstances where the Client appoints a consulting engineer to provide the detailed design of the tunnel it may be appropriate for the GBR to define the geotechnical design parameters utilised in, and construction conditions assumed for, the design of the tunnel linings, but this is not requisite. On schemes for which the Client appoints a contractor to undertake the detailed design in addition to the construction of the tunnel, with a consulting engineer commissioned for outline design only, the GBR would not define geotechnical design parameters, as these would be developed by the Contractor's designer.

3.9 References

Barton, N. R. and Bandis, S. C. (1982). Effects of block size on the shear behaviour of jointed rock. *23rd US Symp. on Rock Mech., Berkeley*, pp. 739–760.

Barton, N. R. and Choubey, V. (1977). The shear strength of rock joints in theory and practice. *Rock Mechanics* **10**, 1–54.

Barton, N. R., Lien, R. and Lunde, J. (1974). Engineering classification of rock masses for the design of tunnel support. *Rock Mechanics* **6**(4), 189–239.

Barton, N. R., By, T. L., Chryssanthakis, L., Tunbridge, L., Kristiansen, J., Loset, F., Bhasin, R. K., Westerdhal, H. and Vik, G. (1992). Comparison of prediction and performance for a 62 m span sports hall in jointed gneiss. *Proc. 4th Int. Rock Mech. and Rock Eng. Conf., Torino*, Paper 17.

Bieniawski, Z. T. (1976). *Rock Mass Classification in Rock Engineering, Exploration for Rock Engineering*. Balkema, Rotterdam, vol. 1, pp. 97–106.

British Standards Institution (1999). *Code of Practice for Site Investigations*. BSI, London, BS 5930.

Carlsson, A., Monaghan, B., Olsson, T., Richards, L. and Ryback, K. (1992). *Support Practice for Large Underground Caverns in Crystalline Rocks – an*

Inventory and Literature Review. Vattenfall Hydro Power Generation, Sweden, Report H 1992/2.

Clayton, C. R. I., Simons, N. E. and Matthews. M. (1982). *Site Investigations*. Granada Technical Books, Manchester.

Construction Industry Research and Information Association (1978). *Tunnelling – Improved Contract Practices*, CIRIA, London, Report 79.

Essex, R. J. (1997). *Geotechnical Baseline Reports for Underground Construction, Guidelines and Practices*, American Society of Civil Engineers, Reston, Virginia.

Glossop, R. (1968). The rise of geotechnology and its influence on engineering practice. *Géotechnique* **18**, pp. 107–150.

Grimstad, E. and Barton, N. (1993). Updating of the Q-System for NMT. *Proc. of the Int. Symp. on Sprayed Concrete, Fagernes, Norway*. Norwegian Concrete Association, 46–66, Balkema, Rotterdam.

Hoek, E. (2000). *Practical Rock Engineering; Hoek's Corner*, Roc Science website: www.rocscience.com.

Hoek, E., Kaiser, P. K. and Bawden, W. F. (1995). *Support of Underground Excavations in Hard Rock.* Balkema, Rotterdam.

Muir Wood, A. (2000). *Tunnelling: Management by Design*. E & F N Spon, London and New York.

Perry, J. and West, G. (1996). *Sources of Information for Site Investigations in Britain*. Transport Research Laboratory, Crowthorne, Report 192 (Revision of TRL Report LR 403).

West, G. and Ewan, V. J. (1982). *Dinorwig Diversion Tunnel*. Transport and Road Research Laboratory, Crowthorne, Report 984.

4 Design life and durability

4.1 Definition

A durable lining is one that performs satisfactorily in the working environment during its anticipated service life. The material used should be such as to maintain its integrity and, if applicable, to protect other embedded materials.

4.2 Design life

Specifying the required life of a lining (see Section 2.3.4) is significant in the design, not only in terms of the predicted loadings but also with respect to long-term durability. Currently there is no guide on how to design a material to meet a specified design life, although the new European Code for Concrete (British Standards Institution, 2003) addresses this problem. This code goes some way to recommending various mix proportions and reinforcement cover for design lives of 50 and 100 years. It can be argued that linings that receive annular grouting between the excavated bore and the extrados of the lining, or are protected by primary linings, for example sprayed concrete, may have increased resistance to any external aggressive agents. Normally, these elements of a lining system are considered to be redundant in terms of design life. This is because reliably assessing whether annulus grouting is complete or assessing the properties or the quality of fast set sprayed concrete with time is generally difficult.

Other issues that need to be considered in relation to design life include the watertightness of a structure and fire-life safety. Both of these will influence the design of any permanent lining.

4.3 Considerations of durability related to tunnel use

Linings may be exposed to many and varied aggressive environments. Durability issues to be addressed will be very dependent not only on the site location and hence the geological environment but also on the use of the tunnel/shaft (see Fig. 4.1).

The standards of material, design and detailing needed to satisfy durability requirements will differ and sometimes conflict. In these cases a compromise must be made to provide the best solution possible based on the available practical technology.

4.4 Considerations of durability related to lining type

4.4.1 Steel/cast-iron linings

Unprotected steel will corrode at a rate that depends upon the temperature, presence of water with reactive ions (from salts and acids) and availability of oxygen. Typically corrosion rates can reach about 0.1 mm/year. If the availability of oxygen is limited, for example at the extrados of a segmental lining, pitting corrosion is likely to occur for which corrosion rates are more difficult to ascertain.

Grey cast-iron segments have been employed as tunnel linings for over a hundred years, with little evidence as yet of serious corrosion. This is because this type of iron contains flakes of carbon that become bound together with the corrosion product to prevent water and, in ventilated tunnels, oxygen from reaching the mass of the metal. Corrosion is therefore stifled. This material is rarely if ever used in modern construction due to the higher strength capacities allowed with SGI linings.

Internal considerations (normal use)	Major influence on durability
Road/rail	Chlorides (de-icing salts)
Sea outfalls	Chlorides (sea water), abrasion
Foul sewers	Acids, sulphates
Storm sewers	Abrasion
Potable water	Watertightness
Cable	Watertightness
Entrance zones	Freeze/thaw, carbonation

Internal considerations (emergency situations)	
Road/rail	Fire, spillages
Cable	Fire

External considerations	
Natural soils	
Clay fissured (mobile water)	Sulphates
Sands/gravels (highly mobile water)	Contaminants from other sources
Brownfield sites (industrial waste) soils	Various contaminants, sulphates, acids

Fig. 4.1 Major factors affecting the durability of tunnel linings

Spheroidal-Graphite cast iron (SGI) contains free carbon in nodules rather than flakes, and although some opinion has it that this will reduce the self-stifling action found in grey irons, one particular observation suggests that this is not necessarily so. A 250 m length of service tunnel was built in 1975 for the Channel Tunnel, and SGI segments were installed at the intersection with the tunnel constructed in 1880. The tunnel was mainly unventilated for the next ten years, by which time saline groundwater had caused corrosion and the intrados appeared dreadfully corroded. The application of some vigorous wire brushing revealed that the depth of corrosion was in reality minimal.

4.4.2 Concrete linings

In situ concrete was first used in the UK at the turn of the century. Precast concrete was introduced at a similar time but it was not used extensively until the 1930s. There is therefore only 70 to 100 years of knowledge of concrete behaviour on which to base the durability design of a concrete lining.

The detailed design, concrete production and placing, applied curing and post curing exposure, and operating environment of the lining all impact upon its durability. Furthermore, concrete is an inherently variable material. In order to specify and design to satisfy durability requirements, assumptions have to be made about the severity of exposure in relation to deleterious agents, as well as the likely variability in performance of the lining material itself. The factors that generally influence the durability of the concrete and those that should be considered in the design and detailing of a tunnel lining include:

- *operational environment*
- *shape and bulk of the concrete*
- *cover to the embedded steel*
- *type of cement*

- *type of aggregate*
- *type and dosage of admixture*
- *cement content and free water/cement ratio*
- *workmanship, for example compaction, finishing, curing*
- *permeability, porosity and diffusivity of the final concrete.*

The geometric shape and bulk of the lining section is important because concrete linings generally have relatively thin walls and are possibly subject to a significant external hydraulic head. Both of these will increase the ingress of aggressive agents into the concrete.

4.5 Design and specification for durability

It has to be accepted that all linings will be subject to some level of corrosion and attack by both the internal and external environment around a tunnel. They will also be affected by fire. Designing for durability is dependent not only on material specification but also on detailing and design of the lining.

4.5.1 Metal linings

Occasionally segments are fabricated from steel, and these should be protected by the application of a protective system. Liner plates formed from pressing sheet steel usually act as a temporary support while an in situ concrete permanent lining is constructed. They are rarely protected from corrosion, but if they are to form a structural part of the lining system they should also be protected by the application of a protective system. Steel sections are often employed as frames for openings and to create small structures such as sumps. In these situations they should be encased in concrete with suitable cover and anti-crack reinforcement. In addition, as the quality of the surrounding concrete might not be of a high order consideration should be given to the application of a protective treatment to such steelwork.

Spheroidal-Graphite cast iron segmental tunnel linings are usually coated internally and externally with a protective paint system. They require the radial joint mating surfaces, and the circumferential joint surfaces, to be machined to ensure good load transfer across the joints and for the formation of caulking and sealing grooves. It is usual to apply a thin coat of protective paint to avoid corrosion between fabrication and erection, but long-term protective coatings are unnecessary as corrosion in such joints is likely to be stifled.

It is suggested that for SGI segmental linings the minimum design thicknesses of the skin and outer flanges should be increased by one millimetre to allow for some corrosion (see Channel Tunnel case history in Chapter 10). If routine inspections give rise to a concern about corrosion it is possible to take action, by means of a cathodic protection system or otherwise, to restrain further deterioration. The chance of having to do this over the normal design lifetime is small.

4.5.1.1 Protective systems
Cast iron segmental linings are easily protected with a coating of bitumen, but this material presents a fire hazard, which is now unacceptable on the interior of the tunnel. A thin layer, up to 200 μm in thickness, of specially formulated paint is now employed; to get the paint to adhere it is necessary to specify the surface preparation. Grit blasting is now normally specified, however, care should be taken in the application of these coatings. The problem of coatings for cast iron is that grit blasting leaves

behind a surface layer of small carbon particles, which prevents the adhesion of materials, originally designed for steelwork, and which is difficult to remove. It is recommended that the designer take advice from specialist materials suppliers who have a proven track record.

Whether steel or cast iron segments are being used, consideration of the ease with which pre-applied coatings can be damaged during handling, erection and subsequent construction activities in the tunnel is needed.

4.5.1.2 Fire resistance Experiences of serious fires in modern tunnels suggest that temperatures at the lining normally average 600–700 °C, but can reach 1300 °C (see Section 4.5.3). It is arguable that fire protection is not needed except where there is a risk of a high-temperature (generally hydrocarbon) fire. It can be difficult to find an acceptable economic solution, but intumescent paint can be employed. This is not very effective in undersea applications. As an alternative an internal lining of polypropylene fibre reinforced concrete might be considered effective.

4.5.2 Concrete linings

All aspects of a lining's behaviour during its design life, both under load and within the environment, should be considered in order to achieve durability. The principle factors that should be considered in the design and detailing are:

- *material(s)*
- *production method*
- *application method (e.g. sprayed concrete)*
- *geological conditions*
- *design life*
- *required performance criteria.*

4.5.2.1 Corrosion The three main aspects of attack that affect the durability of concrete linings are:

- *corrosion of metals*
- *chloride-induced corrosion of embedded metals*
- *carbonation-induced corrosion of embedded metals.*

Corrosion of metals Unprotected steel will corrode at a rate that depends upon temperature, presence of water and availability of oxygen. Exposed metal fittings, either cast in (i.e. a bolt- or grout-socket), or loose (e.g. a bolt), will corrode (see Section 4.5.4). It is impractical to provide a comprehensive protection system to these items and it is now standard practice to eliminate ferrous cast in fittings totally by the use of plastics. Loose fixings such as bolts should always be specified with a coating such as zinc.

Chloride-induced corrosion Corrosion of reinforcement continues to represent the single largest cause of deterioration of reinforced concrete structures. Whenever there are chloride ions in concrete containing embedded metal there is a risk of corrosion. All constituents of concrete may contain some chlorides and the concrete may be contaminated by other external sources, for example de-icing salts and seawater.

Damage to concrete due to reinforcement corrosion will only normally occur when chloride ions, water and oxygen are all present.

Chlorides attack the reinforcement by breaking down the passive layer around the reinforcement. This layer is formed on the surface of the steel as a result of the highly alkaline environment formed by the hydrated cement. The result is the corrosion of the steel, which can take the form of pitting or general corrosion. Pitting corrosion reduces the size of the bar, while general corrosion will result in cracking and spalling of the concrete.

Although chloride ions have no significant effect on the performance of the concrete material itself, certain types of concrete are more vulnerable to attack because the chloride ions then find it easier to penetrate the concrete. The removal of calcium aluminate in sulphate-resistant cement (the component that reacts with external sulphates), results in the final concrete being less resistant to the ingress of chlorides. To reduce the penetration of chloride ions, a dense impermeable concrete is required. The use of corrosion inhibitors does not slow down chloride migration but does enable the steel to tolerate high levels of chloride before corrosion starts.

Current code and standard recommendations to reduce chloride attack are based on the combination of concrete grade (defined by cement content and type, water/cement ratio and strength, that is indirectly related to permeability) and cover to the reinforcement. The grade and cover selected is dependent on the 'exposure condition'. There are also limits set on the total chlorides content of the concrete mix.

Carbonation-induced corrosion In practice, carbonation-induced corrosion is regarded as a minor problem compared with chloride-induced corrosion. Even if carbonation occurs it is chloride-induced corrosion that will generally determine the life of the lining. Carbonated concrete is of lower strength but as carbonation is limited to the extreme outer layer the reduced strength of the concrete section is rarely significant.

Damage to concrete will only normally occur when carbon dioxide, water, oxygen and hydroxides are all present. Carbonation is unlikely to occur on the external faces of tunnels that are constantly under water, whereas some carbonation will occur on the internal faces of tunnels that are generally dry. Carbonation-induced corrosion, however, is unlikely in this situation due to lack of water. Linings that are cyclically wet and dry are the most vulnerable.

When carbon dioxide from the atmosphere diffuses into the concrete, it combines with water forming carbonic acid. This then reacts with the alkali hydroxides forming carbonates. In the presence of free water, calcium carbonate is deposited in the pores. The pH of the pore fluid drops from a value of about 12.6 in the uncarbonated region to 8 in the carbonated region. If this reduction in alkalinity occurs close to the steel, it can cause depassivation. In the presence of water and oxygen corrosion of the reinforcement will then occur.

To reduce the rate of carbonation a dense impermeable concrete is required.

As with chloride-induced corrosion, current code and standard recommendations to reduce carbonation attack are based on the combination of concrete grade and reinforcement cover.

4.5.2.2 Other chemical attack Chemical attack is by direct attack either on the lining material or on any embedded materials, caused by aggressive agents being part of the contents within the tunnel or in the ground in the vicinity of the tunnel. Damage to the material will depend on a number of factors including the concentration and type of chemical in question, and the movement of the groundwater, that is the ease with which the chemicals can be replenished at the surface of the concrete. In this respect static water is generally defined as occurring in ground having a mass permeability of $<10^{-6}$ m/s and mobile water $>10^{-6}$ m/s. The following types of exchange reactions may occur between aggressive fluids and components of the lining material:

- *sulphate attack*
- *acid attack*
- *alkali–silica reaction (ASR).*

Sulphates (conventional and thaumasite reaction) In soil and natural groundwater, sulphates of sodium, potassium, magnesium and calcium are common. Sulphates can also be formed by the oxidation of sulphides, such as pyrite, as a result of natural processes or with the aid of construction process activities. The geological strata most likely to have a substantial sulphate concentration are ancient sedimentary clays. In most other geological deposits only the weathered zone (generally 2 m to 10 m deep) is likely to have a significant quantity of sulphates present. By the same processes, sulphates can be present in contaminated ground. Internal corrosion in concrete sewers will be, in large measure, due to the presence of sulphides and sulphates at certain horizons dependent on the level of sewer utilisation. Elevated temperatures will contribute to this corrosion.

Ammonium sulphate is known to be one of the salts most aggressive to concrete. However, there is no evidence that harmful concentrations occur in natural soils.

Sulphate ions primarily attack the concrete material and not the embedded metals. They are transported into the concrete in water or in unsaturated ground, by diffusion. The attack can sometimes result in expansion and/or loss of strength. Two forms of sulphate attack are known; the conventional type leading to the formation of gypsum and ettringite, and a more recently identified type producing thaumasite. Both may occur together.

Constituents of concrete may contain some sulphates and the concrete may be contaminated by external sources present in the ground in the vicinity of the tunnel or within the tunnel.

Damage to concrete from conventional sulphate reaction will only normally occur when water, sulphates or sulphides are all present. For a thaumasite-producing sulphate reaction, in addition to water and sulphate or sulphides, calcium silicate hydrate needs to be present in the cement matrix, together with calcium carbonate. In addition, the temperature has to be relatively low (generally less than 15 °C).

Conventional sulphate attack occurs when sulphate ions react with calcium hydroxide to form gypsum (calcium sulphate), which in turn reacts with calcium aluminate to form ettringite. Sulphate resisting cements have a low level of calcium aluminate so reducing the extent of the reaction. The formation of gypsum and ettringite results in expansion and disruption of the concrete.

Sulphate attack, which results in the mineral thaumasite, is a reaction between the calcium silicate hydrate, carbonate and sulphate ions. Calcium silicate hydrate forms the main binding agent in Portland cement, so this form of attack weakens the concrete and, in advanced cases, the cement paste matrix is eventually reduced to a mushy, incohesive white mass. Sulphate resisting cements are still vulnerable to this type of attack.

Current code and standard recommendations to reduce sulphate attack are based on the combination of concrete grade. Future code requirements will also consider aggregate type. There are also limits set on the total sulphate content of the concrete mix but, at present, not on aggregates, the recommendations of BRE Digest 363 1996 should be followed for any design.

Acids Acid attack can come from external sources, that are present in the ground in the vicinity of the tunnel, or from within the tunnel. Groundwater may be acidic due to the presence of humic acid (which results from the decay of organic matter), carbonic acid or sulphuric acid. The first two will not produce a pH below 3.5. Residual pockets of sulphuric (natural and pollution), hydrochloric or nitric acid may be found on some sites, particularly those used for industrial waste. All can produce pH values below 3.5. Carbonic acid will also be formed when carbon dioxide dissolves in water.

Concrete subject to the action of highly mobile acidic water is vulnerable to rapid deterioration. Acidic ground waters that are not mobile appear to have little effect on buried concrete.

Acid attack will affect both the lining material and other embedded metals. The action of acids on concrete is to dissolve the cement hydrates and, also in the case of aggregate with high calcium carbonate content, much of the aggregate. In the case of concrete with siliceous gravel, granite or basalt aggregate the surface attack will produce an exposed aggregate finish. Limestone aggregates give a smoother finish. The rate of attack depends more on the rate of movement of the water over the surface and the quality of the concrete, than on the type of cement or aggregate.

Only a very high density, relatively impermeable concrete will be resistant for any period of time without surface protection. Damage to concrete will only normally occur when mobile water conditions are present.

Current code and standard recommendations to reduce acid attack are based on the concrete grade (defined by cement content and type, water/cement ratio and strength). As cement type is not significant in resisting acid attack, future code requirements will put no restrictions on the type used.

Alkali Silica Reaction (ASR) Some aggregates contain particular forms of silica that may be susceptible to attack by alkalis originating from the cement or other sources.

There are limits to the reactive alkali content of the concrete mix, and also to using a combination of aggregates likely to be unreactive. Damage to concrete will only normally occur when there is a high moisture level within the concrete, there is a high reactivity alkali concrete content or another source of reactive alkali, and the aggregate contains an alkali-reactive constituent. Current code and standard recommendations to reduce ASR are based on limiting the reactive alkali content of the concrete mix, the recommendations of BRE 330 1999 should be followed for any design.

4.5.2.3 Physical processes Various mechanical processes including freeze-thaw action, impact, abrasion and cracking can cause concrete damage.

- **Freeze-thaw** Concretes that receive the most severe exposure to freezing and thawing are those which are saturated during freezing weather, such as tunnel portals and shafts.

  Deterioration may occur due to ice formation in saturated concrete. In order for internal stresses to be induced by ice formation, about 90% or more by volume of pores must be filled with water. This is because the increase in volume when water turns to ice is about 8% by volume.

  Air entrainment in concrete can enable concrete to adequately resist certain types of freezing and thawing deterioration, provided that a high quality paste matrix and a frost-resistant aggregate are used.

  Current code and standard recommendations to reduce freeze–thaw attack are based on introducing an air entrainment agent when the concrete is below a certain grade. It should be noted that the inclusion of air will reduce the compressive strength of the concrete.

- **Impact** Adequate behaviour under impact load can generally be achieved by specifying concrete cube compressive strengths together with section size, reinforcement and/or fibre content. Tensile capacity may also be important, particularly for concrete without reinforcement.

- **Abrasion** The effects of abrasion depend on the exact cause of the wear. When specifying concrete for hydraulic abrasion in hydraulic applications, the cube compressive strength of the concrete is the principal controlling factor.

- **Cracking** The control of cracks is a function of the strength of concrete, the cover, the spacing, size and position of reinforcement, and the type and frequency of the induced stress. When specifying concrete cover there is a trade-off between additional protection from external chloride attack to the reinforcement, and reduction in overall strength of the lining.

4.5.3 Protective systems

Adequate behaviour within the environment is achieved by specifying concrete to the best of current practice in workmanship and materials. Protection of concrete surfaces is recommended in codes and standards when the level of aggression from chemicals exceeds a maximum specified limit. Various types of surface protection include coatings, waterproof barriers and a sacrificial layer.

4.5.3.1 Coatings Coatings have changed over the years, with tar and cut-back bitumens being less popular, and replaced by rubberised bitumen emulsions and epoxy resins. The fire hazard associated with bituminous coatings has limited their use to the extrados of the lining in recent times. The risk of damage to coatings during construction operations should be considered.

4.5.3.2 Waterproof barriers The requirements for waterproof barriers are similar to those of coatings. Sheet materials are commonly used, including plastic and bituminous membranes. Again, the use of bituminous materials should be limited to the extrados.

4.5.3.3 Sacrificial layer This involves increasing the thickness of the concrete to absorb all the aggressive chemicals in the sacrificial outer layer. However, use of this measure may not be appropriate in circumstances where the surface of the concrete must remain sound, for example joint surfaces in segmental linings.

4.5.4 Detailing of precast concrete segments

The detailing of the ring plays an important role in the success of the design and performance of the lining throughout its design life. The ring details should be designed with consideration given to casting methods and behaviour in place. Some of the more important considerations are as follows.

- **Eliminate all embedded metallic fittings and fixings, bolt sockets and grout sockets**   (See Section 4.5.2.1.)
- **Thickness and segments size**   Particularly related to handling and transportation.
- **Gasket grooves**   Too small a distance to the edge may result in the enclosing nib breaking under load or when transporting the segment.
- **Joints**   These must be detailed to achieve the required watertightness giving consideration to the type of waterproofing material used.
- **Joint bearings**   Joints must be detailed to achieve adequate bearing area but with reliefs or chamfers to minimise spalling and stripping damage.
- **Overall detail**   Consideration should be given to all tolerances of manufacture and construction.
- **Positioning of fixings**   Embedded fixings/holes should be positioned to allow continuity of reinforcement (when required) while maintaining cover.

4.5.5 Codes and standards

Building Research Establishment (BRE) Digest 330: 1999 (Building Research Establishment, 1999), Building Research Establishment (BRE) Digest 363: 1996 (Building Research Establishment, 1996), BRE Special Digest 1 (Building Research Establishment, 2003) and British Standard BS EN 206-1: 2000 (British Standards Institution, 2003) are the definitive reference points for designing concrete mixes which are supplemented by BS 8110 (British Standards Institution, 1997) and BS 8007 (British Standards Institution, 1987). BS EN 206-1 also references *Eurocode 2: Design of Concrete Structures* (European Commission, 1992).

4.5.5.1 European standards *EN 206 Concrete – Performance, Production and Conformity*, and DD ENV 1992-1-1 (*Eurocode 2: Design of Concrete Structures Part 1*) (British Standards Institution, 2003 and European Commission, 1992).

Within the new European standard EN 206 *Concrete – Performance, Production and Conformity*, durability of concrete will rely on prescriptive specification of minimum grade, minimum binder content and maximum water/binder ratio for a series of defined environmental classes. This standard includes indicative values of specification parameters as it is necessary to cover the wide range of environments and cements used in the EU member states.

Cover to reinforcement is specified in DD ENV 1992-1-1 (*Eurocode 2: Design of Concrete Structures Part 1* – European Commission, 1992).

Tunnel lining design guide. Thomas Telford, London, 2004

4.5.5.2 BRE 330: 1999 This UK Building Research Establishment code (Building Research Establishment, 1999) gives the background to ASR as well as detailed guidance for minimising the risks of ASR and examples of the methods to be used in new construction.

4.5.5.3 Reinforcement BRE 363: 1996 This UK Building Research Establishment code (Building Research Establishment, 1996) discusses the factors responsible for sulphate and acid attack on concrete below ground level and recommends the type of cement and quality of concrete to provide resistance to attack.

4.5.5.4 BRE Special Digest 1 This special digest (Building Research Establishment, 2003) was published following the recent research into the effects of thaumasite on concrete. It replaces BRE Digest 363: 2001. Part 4 is of specific reference to precast concrete tunnel linings.

4.5.5.5 BS 8110/BS 8007 Guidance is given on minimum grade, minimum cement and maximum w/c ratio for different conditions of exposure. Exposure classes are mild, moderate, severe, very severe, most severe and abrasive related to chloride attack, carbonation and freeze–thaw. The relationship between cover of the reinforcement and concrete quality is also given together with crack width (British Standards Institution, 1987a and 1997a).

4.5.5.6 Others Chemically aggressive environments are classified in specialist standards. For information on industrial acids and made up ground, reference may be made to a specialist producer of acid resistant finishes or BS 8204-2 (British Standards Institution, 1999). For silage attack, reference should be made to the UK Ministry of Agriculture, Fisheries and Food.

4.6. Fire resistance

In order to maintain the structural integrity of shafts and tunnels, those with occasion to contain flammable and strong oxidant substances should be designed to withstand the consequences of a fire.

There have been a number of recent high profile fires in tunnels, for example the Channel Tunnel, the Storebaelt running tunnel, the Mont Blanc Tunnel, and the Tauern and Gotthard tunnels. Many of these, particularly the Mont Blanc Tunnel, have resulted in a significant loss of life. In most the integrity of the tunnel structures was not affected although much of the material forming the permanent structural fabric was seriously damaged, and some states of 'near collapse' have been reported.

There are a range of measures and codes of practice available to designers to meet specific requirements. The final choice will depend on the tunnel use and the impact on stability if the lining is seriously damaged. In rock tunnels, stability is less of an issue but in soft ground tunnels, particularly heavily used road tunnels, security of the structure as well as the users is paramount, and this needs to be carefully assessed.

The materials used to line the tunnel and the code of design practice selected should limit the increase in the rise of temperature in the lining, for example to control explosive spalling in concrete or shotcrete, allowing users to evacuate the tunnel safely within a given time frame. The codes of practice that are applied should reflect a worst-case scenario. This is not always easy to establish because although many of the guidelines for tunnel operation may specify that inflammable materials are not to be routed through tunnels, it is impossible to exclude all materials with a high fire rating, for example edible fats, packaging and petrol as vehicle fuel.

4.6.1 Effects of tunnel type and shape

The structural members of tunnels can be divided into two main types: flexural members, for example members of rectangular tunnels, and compression members, for example members of circular tunnels. The form of a tunnel affects its ability to resist fire.

4.6.1.1 Non-circular tunnels In non-circular tunnels or tunnels with non-uniform cross section formed of reinforced concrete, the principal load condition is controlled by considerations of bending. Thus spalling of the soffit and loss of reinforcement in that zone will significantly reduce the capacity of the section.

4.6.1.2 Circular tunnels In circular tunnels, the principal load condition is hoop compression. As such, the reinforcement is generally only detailed for handling requirements and provides only secondary structural support. Exposing the reinforcement, therefore, may not have the same significance and the reduction of capacity may only be governed by the amount and rate of spalling. However, it should be noted that the Channel Tunnel fire resulted in spalling up to 380 mm deep (400 mm-thick units). The spalling was deepest at the centre of the units, and was less pronounced at the four edges where additional reinforcement has been provided.

As a result of the effects of a fire the movement of the tunnel lining, the stiffness, the loss of section and the ground lining interaction will change and should be considered in design.

4.6.2 Types of fire

Typically two types of fire can be distinguished: cellulose fires and hydrocarbon fires.

Cellulose fires refer to the burning of wood, cardboard, paper and similar materials. Such fires generally start slowly and may in enclosed spaces eventually reach temperatures of up to 1000 °C.

The term hydrocarbon fires refers to burning petrochemicals. In contrast to cellulose fires, a hydrocarbon fire usually develops at a considerably higher speed (depending on the volatility of the components) and temperatures of up to 1300 °C may occur.

4.6.3 Lining material behaviour in fire

To understand the response of tunnels to the high temperatures experienced in a fire it is important to have knowledge of the changes that occur to the properties of materials of which they are constructed (see Fig. 4.2).

4.6.3.1 Concrete behaviour in fire Vapour migrates through capillaries to the outer surface. On the heated side the vapour turns to steam, but on the cool side (if exposed) it can condense and appear as 'weeping'. The two main issues for fire design for concrete linings are to prevent explosive spalling (mainly from trapped vapour pressure) and minimise strength loss.

Tunnel lining design guide. Thomas Telford, London, 2004

Chemical	Physical	Mechanical	Thermal
Decomposition[a]	Density	Strain deformation	Conductivity[d]
Charring[a]	Spalling[b]	Strength	
	Softening[c]	Elasticity	
	Melting[c]	Creep (not wood)	
	Deformation		

[a] Wood only.
[b] Concrete and masonry only.
[c] Softening of steel in excess of 800 °C, and in extreme cases fire temperatures can reach 1300 °C at which melting can occur.
[d] Thermal properties influence the rate of heat transfer into the construction.

Fig. 4.2 Material properties significant in a fire

Three types of spalling are generally experienced: aggregate splitting, explosive spalling and sloughing off.

- **Aggregate splitting** is the splitting of aggregates (typically siliceous aggregates) at high temperature due to physical changes in their crystalline structure.
- **Explosive spalling** is related primarily to the permeability of the concrete and its moisture content. Research has shown that such spalling is caused by the development of high vapour pressures trapped within the concrete section, causing cracks to form and concrete layers to be blown away with an explosive force. The main cause is the range of vapour pressure when free water (and, to a lesser extent, water bound physically and chemically) evaporates. This effect causes the free liquid still present in the pores to be pushed into the concrete even further. This results in a saturated zone and the vapour pressures will increase even more. Explosive spalling will occur at various surface temperatures dependent on the type of concrete. Typically 200 °C has been reported.

The effect is more pronounced with a

- *lower permeability concrete*
- *higher moisture content*
- *faster increase in temperature, that is the rate at which the maximum temperature is reached in the concrete*
- *longer exposure time.*

- **Sloughing off** occurs at the surface when the concrete has become weak after prolonged exposure to high temperatures.
- **Explosive spalling** is generally the most destructive of the above.

4.6.3.2 Conductivity The thermal conductivity of concrete depends upon the nature of the aggregate, porosity and moisture content. As the water is driven from the concrete in a fire, the conductivity of dry concrete is probably more relevant. Various concretes have been examined and it has been found that for dense concretes, conductivity decreases with increasing temperature but for lightweight concretes the decrease is nominal.

4.6.3.3 Cement paste Heating causes removal of water, which causes shrinkage and changes in density, and modifies bonding forces between the minute crystals that comprise the cement gel. This then affects concrete stiffness, strength and dimension. A temperature above 300 °C leads to loss of strength and the chemical changes that are experienced lead to a change in colour.

4.6.3.4 Aggregates Calcareous aggregate, for example limestone, neither changes its physical structure nor undergoes any sudden expansion during heating except at very high temperatures. Siliceous aggregate, for example flint, gravel, granite, are known to expand at high temperatures, which in turn may lead to spalling of the concrete by aggregate splitting. Physical changes also take place, for example quartz at 575 °C.

4.6.3.5 Reinforcement The principal types of reinforcing material used in the UK are specified in BS 4449 (British Standards Institution, 1997).

As steel is heated the tensile strength, yield stress and modulus of elasticity are reduced. The ultimate strength is therefore higher than the yield strength when the steel is in the cold state and hence more deformation will occur before it breaks.

The yield stress of hot-rolled reinforcement is reduced at temperatures above 300 °C. Typically, the yield stress is reduced to around 50% of its value at normal temperatures on heating in the range 550–600 °C. The original yield stress is almost completely recovered when cooling in the range 500–600 °C, but on cooling from 800 °C a reduction of 30% occurs for cold-worked reinforcement and 5% for hot-rolled reinforcement.

4.6.4 Codes and other standards

The fire resistance of elements of construction in the UK is determined by conducting laboratory tests following the procedure laid down in British Standard BS 476 Part 8: 1982 (British Standards Institution, 1987b). Most countries have standards for this purpose. The international specification ISO 834 (International Standards Organisation, 1975) was introduced to harmonise all standards. Fires in underground spaces are still generally unclassified. Nationally and internationally different curves have been developed to simulate fires in tests.

4.6.4.1 BS 476 This uses a typical building fire based on a cellulose fire. This fire profile has a slow temperature rise up to 1000 °C over 120 minutes (British Standards Institution, 1987b).

4.6.4.2 ISO 834 Similar to the BS 476 fire curve, this International Standard uses a typical building fire based on a cellulose fire. This fire profile has a slow temperature rise up to 1100 °C (International Standards Organisation, 1975).

4.6.4.3 Eurocode 1 The following nominal fire curves are included (European Committee for Standardisation, 1991).

Standard fire (cellulose) in which

$$T = 20 + 345 \log_{10}(8t + 1)$$

Hydrocarbon fire in which

$$T = 20 + 1080(1 - 0.325 \, e^{(-0.167t)} - 0.675 \, e^{(-2.5t)})$$

where t is time (minutes) and T is temperature (°C).

In the Netherlands and Germany different curves for hydrocarbon fires are applied for tunnels. The fire curve is based on the scenario of an accident with a tanker carrying 4500 l of petrol exploding in a 12 m-diameter tunnel, producing a fire load of 300 MW and possibly causing a fire lasting two hours.

4.6.4.4 RWS (Rijkswaterstaat) This fire profile (Rijkswaterstaat, 1995) from the Netherlands Department for Public Works has a rapid temperature rise to 1200 °C, peaking at 1350 °C. The complete fire duration is assumed to be two hours. The Rijkswaterstaat specifies a maximum permissible surface temperature for concrete of 380 °C, and for steel of 500 °C.

4.6.4.5 RABT In Germany this fire profile (RABT, 1997) has a rapid temperature rise up to 1200 °C for 1 hour then diminishes to 0 °C over a further 1 hour 40 minutes.

4.6.4.6 Miscellaneous In Switzerland the maximum permissible concrete surface temperature is 250 °C. To meet these criteria an external protective coating is generally required (see Section 4.6.6).

4.6.5 Design for fire

British Standard BS 8110 Part 2 Section 4 (British Standards Institution, 1985) sets out three ways to determine the fire resistance of reinforced concrete members: tabulated data, fire test and fire engineering calculations.

In all cases the size and shape of the element together with the minimum thickness and cover to reinforcement influence the fire resistance. Allowance is also made for the moisture content of the concrete, the type of concrete and aggregate used and whether any protection is provided.

4.6.6 Fire protection

Two basic options for fire protection are available.

- **Protect externally** Protect the concrete against a fast rise in temperature by means of a fire resistant isolation.
- **Protect internally** Protect the concrete against the formation of high vapour stresses.

Both measures if applied as part of the concrete placement will have the advantage of being effective from the construction stage.

4.6.6.1 External protection A degree of protection can be given against relatively low temperature fires by the application of external systems in the form of boarding or spray-applied coatings. Detailed performance criteria and advice should be obtained from specialist suppliers.

4.6.6.2 Internal protection Polypropylene fibres can be added to the concrete mix. These fibres melt at approximately 160 °C and form micro-channels, which can prevent or diminish the occurrence of high vapour pressures, and hence reduce a tendency to spalling.

4.6.7 Fire repair

After a fire it will always be necessary to establish the residual strength of the concrete and reinforcement.

4.6.7.1 State of concrete Tests to determine the strength of concrete after a fire include:

- **Colour change** When concrete is heated a change in colour occurs which is irreversible and allows an estimation to be made of the temperature to which it has been heated
- **Cut cores and test**
- **Other non-destructive methods**.

4.6.7.2 Condition of reinforcement The strength of the reinforcement can be estimated according to an assessment of the maximum temperature to which it has been subjected, and by removing samples and performing mechanical tests.

4.7 Waterproofing

The strategy put in place for achieving the functional and operational requirements for a project will depend on the design requirements. Guidance relating to watertightness and permissible levels of leakage into sub-surface facilities has been presented by the International Tunnelling Association, Haack (1991). In the absence of any other criteria this provides a reasonable basis for an initial evaluation of design requirements, a useful summary of the effects of water ingress on different types of lining, and the most appropriate repair methods. It also serves as a reminder of the benefits of waterproofing systems.

Reference should also be made to the *Civil Engineering Specification for the Water Industry* (*CESWI 5th edition*) (Institution of Civil Engineers, 1997). To achieve control over water inflows and seepage into a tunnel there are a number of products available including membranes, gaskets, injected water stops, and annular and ground grouting.

4.7.1 Membranes

Membranes are available in two main types:

- **Sheet membrane** Sheet membranes that include materials such as PVC (polyvinylchloride), HDPE (high density polyethylene) and PO (polyolefin)
- **Spray-on membrane** Spray-on membranes are a recent innovation and essentially consist of either cement- or rubber-based compounds (although manufacturers are constantly developing new products).

There is a lack of experience with spray-on membranes and therefore most of the following comments and guidelines apply to sheet membranes.

4.7.1.1 Drained systems On most projects a drained water management system that deflects water from around the tunnel via a watertight membrane and geotextile fleece into an invert drainage system is adequate. This type of system limits the development of hydrostatic pressures on the permanent lining and for most practical purposes these need not be taken into account in the design. A detailed evaluation of the long term effectiveness of such a system is required as although they are often satisfactory they can sometimes cause problems. For example the Melbourne City Tunnel, a three lane road tunnel, originally intended to be designed as a drained tunnel was in the event tanked as it was found that due to the local chemistry of the ground and water, the inflows, when oxygenated, would have blocked the originally designed drainage system. Drainage of tunnels in certain ground conditions with relatively low seepage conditions, and in remote areas may be acceptable. But in others, especially in urban areas, drained systems are generally totally unacceptable due to environmental considerations.

Drained systems generally require only a single sheet membrane and this should be supplied with a signal layer that allows defects to be identified and repaired before placement of the final lining. Welding of the seams is a particular problem and historical data, for example Lyon Metro where as many as 50% of the welds were considered defective, suggest that quality control is very important in providing a continuous umbrella. The systems and installation techniques have improved to overcome some of the construction issues but there is little doubt that the best results are achieved using specialist sub-contractors.

The geotextile is laid against the primary lining or rock in order to channel the water to the invert drain. This also protects the membrane from puncture.

4.7.1.2 Watertight systems Full-perimeter membranes present greater difficulties than drained systems. Experience indicates that these waterproofing systems can be effective but the evidence from most projects is that it is extremely difficult to obtain a fully watertight system. A number of devices are available to improve confidence that leakage through the membrane will be limited, and these include double-sheeted systems that can be separated into compartments and tested and grouted to provide better quality control during installation. Such systems also provide greater resistance to puncturing. Geotextile layers are often laid against the primary lining or rock to protect the membrane from puncture during installation.

These are becoming common practice in some countries, including Switzerland, Sweden and Germany, where protecting the environment by prevention of groundwater lowering is a priority. The issue of leakage assumes greater importance when full hydrostatic pressures are included in the lining design and as a load case; this is referred to in Section 5.3.3.

In terms of design of the membrane system it is still important to recognise that the technology for producing a guaranteed fully watertight system is not yet available. Therefore, even watertight systems need to anticipate leakage through the lining and recent designs, for example the Grozier Tunnel, Switzerland, has included a simple drainage layer inside a PVC membrane. This practice is recommended, particularly at higher pressures.

Precedent practice indicates that the designed maximum hydrostatic pressures can approach 70 m head. In principle, there is no limit to the design pressures providing that the lining is capable of carrying the full load and can be constructed. In practice, such designs become uneconomical and therefore it is unlikely that design pressures will substantially exceed current practice unless dictated by environmental considerations and factors other than economics.

The choice of membrane depends on a number of factors and these are provided for guidance in Fig. 4.3. These are for reference only but there has been a substantial amount of research and testing to justify their use.

There is insufficient research and information at present to make similar recommendations for spray-on membranes. General

Parameter	Criteria	PVC	HDPE	PO
Temperature	<23 °C	✓	✓	✓
	>23 °C		✓	✓
Pressure	<1 MPa	✓	✓	✓
	>1 MPa		✓	✓
Groundwater chemistry	Sulphates >1500 mg/l		✓	✓
	Sodium >100 mg/l		✓	✓
	Chlorides >50 mg/l		✓	✓
Groundwater flows	Low	✓	✓	✓
	Medium	✓	✓	✓
	High		✓	✓

Fig. 4.3 Parameters influencing selection of type of sheet membrane

considerations suggest that they are difficult to apply under wet conditions and in terms of the criteria in Figure 4.3 would only be used as an alternative to PVC and only as part of a drained system.

Codes of practice, specifications and guidelines have been developed in countries such as Switzerland and Germany for sheet membranes, and associated quality control procedures, to avoid leakage.

For the future, reference is made here only to the draft prEN 13491 *Required Characteristics for Geomembranes and Geomembrane-related products used in Tunnels and Underground Structures* (European Commission, 1999).

4.7.2 Gaskets

Gaskets are available in two main types:

- **EPDM**   EPDM or neoprene compression gaskets fitted around individual precast concrete or SGI segments.
- **Hydrophilic**   Hydrophilic seals made from specially impregnated rubbers or specially formulated bentonitic-based compounds that swell when in contact with water.

Both EPDM (Ethylene Polythene Diene Monomer) compression gaskets and hydrophilic seals are commonly specified to provide waterproof joints between adjacent segments in a precast concrete tunnel lining. These are not for waterproofing the concrete itself, but to prevent water flow through potential apertures. The usual practice is to employ a single EPDM gasket or single strip of hydrophilic seal, but if this is breached it can be difficult to locate the weakness. A double-seal arrangement has been used or gaskets incorporating through thickness barriers (e.g. the Elbe Tunnel, Hamburg). Alternatively a second preformed sealing groove with injection points has been provided as a means of remedial sealing (see Section 4.7.3) (e.g. London Heathrow Airport T5, the Heathrow Express tunnels and the Channel Tunnel Rail Link).

The long-term durability and deterioration of the performance of the seal due to creep and stress-relief should also be taken into account. The likely fluctuation in water level will dictate the type of gasket to be employed. Hydrophilic seals may deteriorate if repeatedly wetted and dried. Performance can also be affected by the salinity or chemical content of the groundwater. Different hydrophilic seals are required for saline and fresh water.

The performance of these seals with respect to water pressure, gasket compression characteristics and joint gap tolerances is an important part of the lining design. The specification of the type and performance of the sealing system to be used must be carried

Tunnel lining design guide. Thomas Telford, London, 2004

out in conjunction with expert suppliers. The exact system should be determined with the Contractor as it depends on the type of TBM to be used and the detailed design of the erection equipment.

Gasket compression forces have an important influence on the joint design as they can require large forces to close the joints and then hold the joint together while erection continues. The design of the fixings between segments and their performance under load is an integral part of the gaskets' performance. All stages of the erection process must be considered.

Positioning and size of compression gaskets or hydrophilic sealing systems can significantly reduce the cross-sectional areas of joints available for the transfer of compression loads and must be taken into account. Relief behind the gasket can help reduce damage caused by gasket compression by providing a void for the gasket to flow into thereby preventing the gasket from becoming over compressed and behaving in an hydraulic manner. The joint connection, strength, number and position must be designed to ensure and maintain adequate gasket performance.

4.7.3 Injectable gaskets and seals

These waterproofing systems comprise resin grout injected, open pored, foam-filled, channel sections that can be placed to a predetermined design in the joints of concrete structures. They form the construction, expansion or contraction joints and possible surface penetrations within the structure and can be designed to cater for the calculated long-term settlement and movement of the structure.

4.7.3.1 Design   The detailing and spacing of the joints shall:

- *ensure the optimum conditions for both construction methods, planned crack inducement and long-term waterproofing requirements*
- *ensure a maximum acceptable average leakage is achieved*
- *be capable of withstanding medium to high water pressures and be unaffected by extended drying and wetting cycles*
- *be capable of effectively sealing joints between different base materials such as in situ concrete and precast concrete, steel liners and/or shotcrete*
- *be capable of being adapted to different shapes and interfaces independent of the method of construction.*

The void space within the components is of sufficient capacity to transport a re-injectable, resin grout efficiently along the void and between the formed joint surfaces.

4.7.3.2 Materials   Resin grouts are non-toxic, dual-component, metacrylamide resins that ensure protection of the reinforcing steel, and are capable of a controlled hydrophylic expansion of between 10 and 15% when in contact with water, thus providing a positive over-pressure against the sides of the joint. Resins should have a low viscosity to ensure the penetration of the smallest hair-line cracks.

4.7.4 Grouting for leakage prevention

Leaks through SCL and in situ linings can be prevented by grouting the surrounding ground (see Section 3.7.3.1) with cementitious or chemical mixes. Close monitoring of the injection pressures and volumes together with any movements of the surrounding soil/rock structure and the lining are necessary to prevent the lining being over stressed and possible damage.

Primary or secondary grouting of the annulus of a segmental ring may provide some short-term relief to leaks but will not provide a long-term seal against water penetration because the grout will crack as the ring deforms. However, even with segmental linings, grouting the surrounding ground can have a beneficial effect, but great care should be taken in order to prevent distortion or possible damage to the ring, not to grout at an excessive pressure close to the lining.

4.8 References

Building Research Establishment (1996). *Sulphate and acid resistance of concrete in the ground*. BRE, Garston, Watford, BRE Digest 363.

Building Research Establishment (1999). *Alkali–silica Reaction in Concrete: Detailed Guidance for New Construction*. BRE, Garston, Watford, BRE Digest 330 part 2.

Building Research Establishment (2003). *Concrete in Aggressive Ground, Parts 1 to 4*. BRE, Garston, Watford, BRE Special Digest 1.

British Standards Institution (1985). *Structural Use of Concrete: Code of Practice for Special Circumstances*. BSI, London, BS 8110: Part 2.

British Standards Institution (1987a). *Code of Practice for Design of Concrete Structures for Retaining Aqueous Liquids*. BSI, London, BS 8007.

British Standards Institution (1987b). *Fire Tests on Building Materials and Structures*. BSI, London, BS 476.

British Standards Institution (1997a). *Structural Use of Concrete: Code of Practice for Design and Construction*. BSI, London, BS 8110-1.

British Standards Institution (1997b). *Specification for Carbon Steel Bars for the Reinforcement of Concrete*. BSI, London, BS 4449.

British Standards Institution (1999). *Screeds, Bases and In-situ Floorings: Concrete Wearing Surfaces – Code of Practice*. BSI, London, BS 8204-2.

British Standards Institution (2003). *Concrete. Part 1: Specification, Performance, Production and Conformity*. BSI, London, BS EN 206-1.

Directoriaat-General Rijkswaterstaat (DGR) (1995). *Richtlijnen vervoer gevaarlijke stoffen door tunnels gelegen in autosnelwegen*. DGR, Utrecht, Netherlands.

European Commission (1992). *Eurocode 2: Design of Concrete Structures Part 1*. Draft directive DD ENV 1992-1-1.

European Commission (1999). *Required Characteristics for Geomembranes and Geomembrane-related Products used in Tunnels and Underground Structures*. Draft prEN 13491.

European Committee for Standardisation (1991). *Eurocode 1: Basis of Design and Actions on Structures. Part 2: Actions on Structures Exposed to Fire*. Brussels, Belgium.

Haack, A. (1991). Water leakages in subsurface facilities: required watertightness, contractural matters, and methods of redevelopment. *Tunnelling and Underground Space Technology*, **6**, 3, 273–282.

Institution of Civil Engineers (1997). *Civil Engineering Specification for the Water Industry 5th edition*. ICE, London.

International Standards Organisation (1975). *Fire Resistance Test – Elements of Building Construction*. ISO 834.

Richtlinien fuer die Ausstattung und den Betrieb von Strassentunneln (RABT) (1997). *Forschungsgesellschaft fur Strassen-und Verkehrswesen*.

5 Design considerations

5.1 Introduction

5.1.1 Objectives

The aim of this chapter is to provide tunnelling engineers with a framework for managing the engineering design process. Guidelines are provided covering most technical issues and it is intended that, combined with the engineer's judgement and experience, they give valuable assistance in the preparation of concept and detailed designs that meet project performance and safety requirements.

In 1994 the UK health and safety legislation was altered, placing specific duties on designers, which required them to identify and take account of foreseeable risks that could occur during the construction of their designs. *The Construction (Design and Management) Regulations 1994* (HMSO, 1994) and the associated Approved Code of Practice (revised in 2001) defined 'the designer' in very broad terms and included any person or organisation producing a drawing, a set of design details, a specification or a bill of quantities. Persons engaged on these activities in relation to tunnelling work need to be fully conversant with the legal requirements. Involving the designer in health and safety issues is a positive step with a range of potential benefits and is based on a risk assessment and risk management approach. The first principle of this is that having identified a potential risk the designer must then give consideration to its elimination. Close cooperation between those with design skills and those with construction experience is a prerequisite for the best solutions. Anderson and Lance (1997) give a summary of the benefits from this risk assessment approach.

It is inevitable that experience of construction under a wide range of conditions is necessary for fully qualified tunnel designers, especially when considering constructability and health and safety issues. Combined with appropriate use of current technology, particularly numerical methods of analysis, designers have a powerful means of understanding ground/support interaction and assessing the compatibility of ground behaviour and support systems in relation to the proposed construction method. In the context of the risk management process, this is valuable in mitigating and reducing risks to the maximum extent possible. The use of such methods is demanding, both in terms of understanding the limitations of the method and, as discussed in Section 3.6, the need for realistic input parameters.

Chapter 6 provides guidance on methods of analysis.

5.1.2 Tunnel design practice

Tunnelling as an engineering discipline is unique. Traditionally it has relied heavily on experience and most lining designers came from a structural engineering background. The natural approach was to estimate, as accurately as possible, the magnitude and distribution of loads applied to a tunnel support system and then detail the lining to carry the loads.

As a means of estimating these loads, Terzaghi (1946) introduced his classification system for the design of support systems for rock

---

tunnels in 1946. It marked the first time that observations made on the basis of direct experience and knowledge of precedent practice were formulated into an empirical approach that predicted loads for general rock-mass conditions.

More sophisticated empirical approaches emerged in the 1970s for use in rock tunnels; for example, those of Bieniawski (1984) and Barton (1976). The support predictions are generally conservative. However, such methods should be used with care since the lining design has to be tailored to meet the specific requirements of each project. The work of engineers such as Szechy (1967), Rabcewicz (1969), Muir Wood (1975), Curtis (1976), Hoek and Brown (1980) and Einstein and Schwartz (1979), have been instrumental in developing a greater awareness of the factors that govern good design practice. Recently, the International Tunnelling Association (ITA) (1988) has given general guidelines for tunnel design, and specific guidance for segmental tunnel linings (AFTES, 1999). The *Tunnelling Engineering Handbook* (2nd ed) (Bickel *et al.*, 2002) also provides useful guidance.

A wide variety of support systems are available and the lining design will depend on the choice of construction method, including the nature of any temporary support. For segmental linings the division between temporary and permanent support measures is clear. As indicated in Section 1.4, 'temporary' and 'permanent' supports are not necessarily synonymous with 'primary' and 'secondary' linings in sprayed concrete lined (SCL) tunnels. The primary lining is often the responsibility of the designer rather than the Contractor. Some of the factors contributing to this are:

- *increasing use of design-and-build contracts coupled with closer integration of the design and construction process*
- *increasing use of ground improvement techniques*
- *better understanding of the need for the control of deformations during construction*
- *increasing use of sequential excavation methods for a wide range of ground conditions*
- *more frequent use of additional support measures such as dowels or sprayed concrete to control face stability and deformations ahead of the advancing face*
- *improved materials and additives that can target specific design requirements*
- *greater emphasis on health and safety issues.*

5.1.3 Fundamental design concepts

There will always be exceptions to general rules. Therefore an understanding of the conceptual side of tunnel design is necessary for any designer, as well as a broad understanding of construction methods. Tunnelling involves many difficult issues, including:

- *the relationship between volume losses (and the strain in the ground) and the pressures acting on tunnel linings*
- *the load transfer process for large span excavations constructed using multiple headings*
- *design approaches for extreme environments (e.g. squeezing rock)*
- *the benefits derived from the early age behaviour of sprayed concrete in limiting stress concentrations*
- *whether current codes of practice are appropriate for the design of tunnel linings.*

5.2 Engineering design process

5.2.1 Design management

In planning the approach to design it is useful to look at the stages involved in simple terms (Fig. 5.1). Typically most projects pass through concept, detailed design, construction and post-construction stages. Of these, the concept design is the most important.

Within concept design, the use of brainstorming sessions with key staff is particularly effective early in a project. They should be used to identify and categorise the main hazards and opportunities. The composition of the team is important. Many client organisations implement strategies to include contractors as well as designers. This recognises that key decisions often need to be made early (e.g. the procurement of tunnel boring machines (TBMs)), and this usually impacts directly on selection of the tunnel lining, Isaksson (2002). As documenting the outcome of these sessions is useful for the designer during detailed design.

Attention should also be paid to the organisation of and lines of communication on a project. Good management control is necessary where the method of construction relies on interpreting the ground conditions and/or making adjustments to suit actual performance. Traditionally this has been applied on rock tunnelling projects, or in soft-ground tunnelling using sequential excavation methods such as the New Austrian Tunnelling Method, but it also holds true for complex TBM tunnels, such as those involving earth pressure balance (EPB) TBMs. If an 'observational approach' is selected the consequences of this must be reflected in the construction management system.

Realistically, not all risks can be avoided on a project and the detailed design should be developed on the basis of reducing risks to 'As Low As Reasonably Practical', that is, the ALARP principle.

Typically a design should:

- *ensure that there are no inherent structural defects which could lead to catastrophic failure*
- *have deformations contained so that the structure is not over-stressed*
- *have material strengths of load-bearing elements maintained for the design life*
- *have support system capacity that is not exceeded if time-dependent behaviour or other long-term effects are experienced*
- *provide durability throughout the design life*
- *control groundwater inflows and/or outflows from tunnels under internal pressure*
- *evaluate constructibility requirements, for example ensure that erectors for segmental linings will not cause damage to the lining (see Section 5.4.1.4).*

The methods of analysis that can be employed are discussed in detail in Chapter 6. The processes that should be applied during design include:

- *an assessment of the ground behaviour and the proportion of the strength of the ground that is mobilised in response to a given excavation and support sequence*
- *an assessment of the deformations and strains for the moderately conservative and worst credible lining load cases*
- *assessing geometries and lining construction sequences to ensure that adequate load paths exist to prevent over-stressing, particularly at the face*

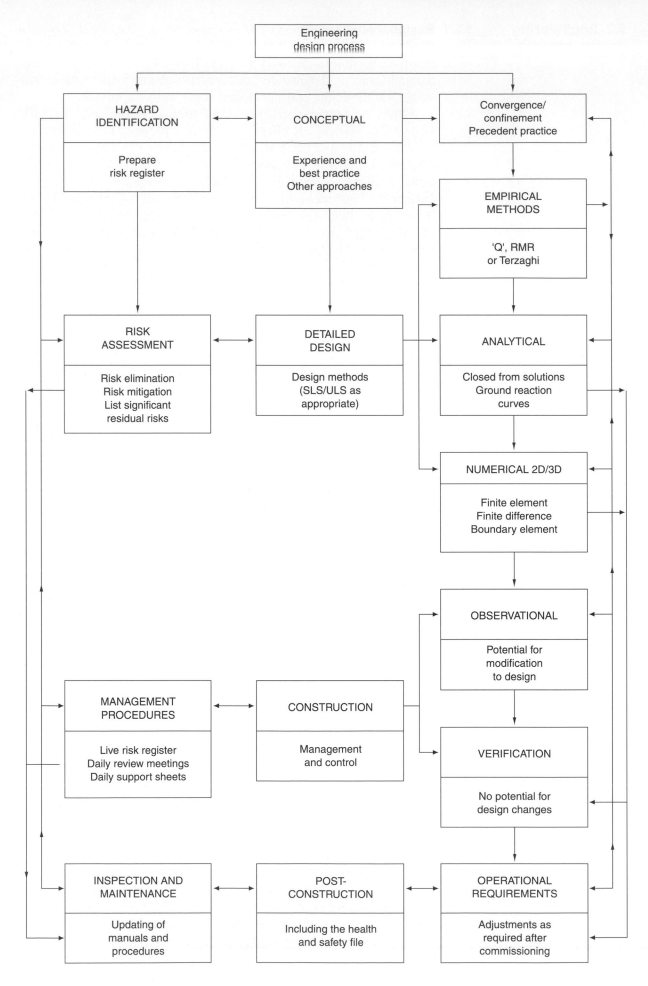

Fig. 5.1 Design stages and approaches

- *a demonstration that there is an adequate factor of safety for each load case.*

The load cases that should be designed for are discussed in more detail later in this chapter.

Since the models and behaviour patterns are complex, the results of the analysis have to be interpreted and evaluated in terms of the consequences of failure to the project. This is part of the risk management process. In applying the ALARP principle, designers will determine what measures should be taken in advance of and during construction to control risks. This usually includes installing instrumentation to monitor the performance of both the ground and lining during construction as well as, sometimes, that of adjacent structures. As noted earlier, if the construction phase involves an element of interpretation, a Client may be advised that the designer be represented on site to ensure that:

- *the construction is in accordance with the design intent, for example the installed support is compatible with the ground conditions*
- *variations offered by the Contractor are compatible with design assumptions and criteria including those related to Health and Safety*
- *unexpected events can be dealt with before they become critical.*

If there are unexpected events, teamwork involving designers and constructors should provide a suitable and timely response. The lessons from the past indicate that this is a small cost compared to the costs (direct and indirect) resulting from a tunnel collapse.

In the post-construction period there may be a requirement for long-term monitoring, for example on rail and road tunnels where public health and safety is dependent on the performance of the civil structures. For more details see Section 8.4.

5.3 Design considerations

5.3.1 Ground/support interaction

The complex interaction between the tunnel support system and ground is strongly influenced by the construction method. Increasingly, this process can be modelled explicitly using numerical methods of analysis (see Chapter 6). However, this is difficult to do and considerable judgement and experience is required in interpreting the results. To check whether designs are sufficiently robust certain principles of ground–support interaction should be considered.

5.3.1.1 Stand-up time Almost all ground properties are time-dependent and strain controlled, both in the short and long term. In the short term, the concept of stand-up time is used as a practical means of indicating the sensitivity of the ground to imposed stress changes. This influences both the support requirements and geometry of tunnel linings. Consolidation, swelling and squeezing may be significant in the longer term.

5.3.1.2 Pressures The pressures acting on a tunnel are usually not well defined. They depend on the excavations and support sequence for any given set of ground conditions. In general, as much as 30–50% of the deformation experienced during construction will occur ahead of the face. Depending on the extent of this deformation ahead of the face, the pressures acting on the lining will be significantly less than predicted from a simplistic two-dimensional 'wished-in-place analysis' that assumes no stress relief. This is an important benefit.

5.3.1.3 Deformation In most cases, the bending strength and stiffness of structural linings are small compared to those of the surrounding ground. The ground properties therefore dictate the distortional deformations and changing the properties of the lining usually will not significantly alter this deformation. On the other hand a completed lining resists uniform hoop deformation well. In general what is required ideally, if it is possible, is a confined flexible lining that can redistribute stresses efficiently without significant loss of load-bearing capacity. This applies to both segmental and sprayed-concrete linings.

5.3.1.4 Immediate support Ground support, when required, usually needs to be installed as close to the face as possible because a large proportion of the deformation occurs ahead of the face. While there are a number of well-understood exceptions to this (notably tunnelling in highly stressed rock) if the installation of support is delayed, it could result in progressive deformations. Consequently there will be a need for additional support to control those deformations or the loosening and dead loads acting on the lining. This is not good design or construction practice and could lead to instability.

The ground/support interaction process has been modelled using Ground Reaction Curves (GRCs), which use a simple analytical approach to relate deformation in the ground with the supporting pressure acting outwards on the extrados of an excavation, resisting the inward deformation (see Fig. 5.2).

Considering GRCs one may note the following:

* **Elastic component** There is generally an elastic component that will occur regardless of when the support is installed.
* **Deformation** The amount of deformation related to non-linear behaviour will be a function of the timing of installation of the support.

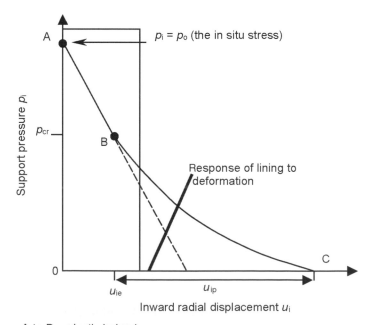

Fig. 5.2 Typical Ground Reaction Curve (GRC) showing an elastic and non-linear ground response for a rock tunnel (after Hoek and Brown, 1980)

A to B = elastic behaviour
B to C = plastic behaviour
u_{ie} = elastic displacement
u_{ip} = plastic displacement
p_{cr} = critical support pressure defined by initiation of plastic failure
    of the rock surrounding the tunnel

Tunnel lining design guide. Thomas Telford, London, 2004

- **Support time** In theory support can be installed at different times to control strains and, as long as the support resistance prevents the onset of loosening, the system will achieve a new state of equilibrium. During construction, monitoring ground and lining movements can check this.
- **Full support** In practice, if the ground requires support, the full support should be installed without delay, unless there are special considerations related to time-dependent behaviour or high ground stresses.
- **Loosening concept** Although it is easy to understand the potential for loosening conceptually, the simple theory behind GRCs does not model this. Furthermore, ground–support interaction is not modelled so the point where the ground reaction curve and the support response line meet only represents an estimate of the required support pressure. This is provided by pressure acting on the lining and the associated ground deformation. The real situation is more complex.

More detailed guidance on the methods of analysis to examine ground–support interaction is given in Chapter 6.

5.3.1.5 Coping with variability in the ground Very often it is not possible to predict accurately the geomechanical properties of the materials and their probable behaviour (see Section 3.5). Therefore the level of effort at the site investigation stage will depend on the likely impact of any potential variations on construction. The use of full-face mechanised tunnelling methods mitigates risks during construction. Segmental linings installed behind the cutterhead are able to tolerate wide variations in ground conditions. Where TBMs are either uneconomical or impractical, potential variations are addressed during construction by ensuring that a range of suitable support systems is available, and by accepting use of an observational approach (Peck, 1969; Nicholson *et al.*, 1999). In relatively soft ground, and particularly for high-risk areas such as shallow urban environments, this type of approach is less acceptable. Here a higher level of effort is required to evaluate the ground conditions and prove robust design (see Section 5.8).

5.3.2 Time-related behaviour

Stress readjustments occur in response to the formation of a tunnel and the effects on the radial and tangential stress distributions are illustrated in Fig. 5.3. As the plastic zone develops the peak tangential stress is transferred away from the edge of the excavation. The simple diagram represents a complex process that occurs over a period of time because the stress readjustment is a three-dimensional process and is related to progress of a heading (see Fig. 5.4). Plane strain conditions and equilibrium are established only after a heading has advanced approximately two diameters beyond the point in question.

This has been well documented and the process is well understood for some geomaterials. For example, considering a shallow tunnel in London Clay, small movements are detected typically up to two diameters ahead of the face and these increase sharply about half a diameter from the face (Van der Berg, 1999). Movements continue at a reasonably constant rate until the closed lining takes full effect and the movements soon cease. This is about one diameter behind ring closure for a SCL tunnel.

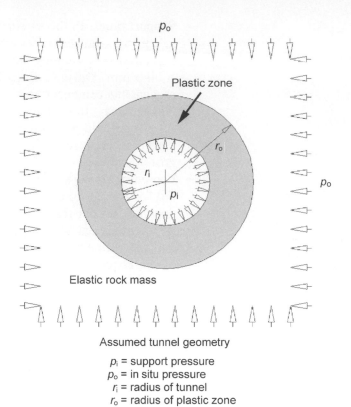

Fig. 5.3 Stress around an opening in an elasto–plastic medium (from Hoek and Brown, 1980)

Assumed tunnel geometry

p_i = support pressure
p_o = in situ pressure
r_i = radius of tunnel
r_o = radius of plastic zone

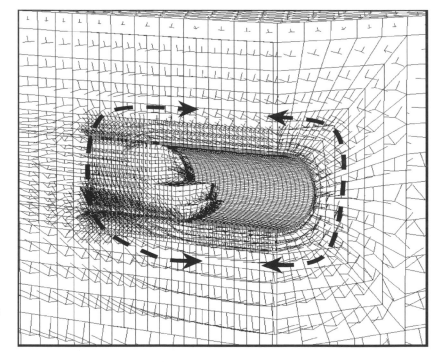

'Arching' of the ground stresses

Fig. 5.4 Three-dimensional stress redistribution around a heading: plot of principal stress directions in a 3D numerical model

5.3.2.1 Timely ring closure   It is apparent from Fig. 5.5 that the steep gradient of the curve will continue until such time as the invert is closed. If closure is delayed, both surface settlement and the associated volume losses will be higher. It is therefore necessary in soft ground to specify that closure should be within less than one diameter. Volume losses of less than 1.0% can be achieved in soft ground with sprayed concrete linings.

Similar values are achievable for segmentally lined TBM-driven tunnels. Higher volume losses occur if there are delays in excavation and grouting of the rings or if the face is not adequately supported. Fully grouting the annulus around the rings as they emerge from the tail-skin, and active support of the face, for example as in

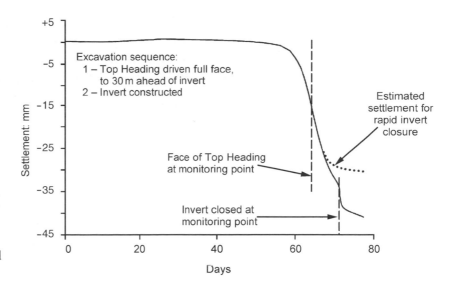

Fig. 5.5 Surface settlement on the centreline above a shallow SCL tunnel in London Clay, showing effect of delayed invert closure (after Deane and Bassett, 1995)

EPB TBMs, can reduce volume losses to less than 0.5% in some ground conditions.

Speed of closure of the invert is not applicable to most rock tunnels where a solely elastic response is anticipated. There will be a finite amount of deformation and depending on the stiffness of the rock mass and the depth of cover, volume losses of the order of 0.05–0.5% will occur.

5.3.2.2 Squeezing and swelling These represent special cases where time-dependent behaviour imposes additional constraints on the excavation and support sequence, and much higher volume losses have to be allowed for. This particularly applies to Alpine tunnels where residual tectonic stresses are locked into the rock mass and release occurs over relatively long periods of time, for example several months. The Arlberg Tunnel (see Fig. 5.6) was a classic example of squeezing ground. Volume losses of up to 10% required a very flexible support system to control deformations and maintain stability. This can be achieved by means of yielding arch supports or by leaving longitudinal slots in the sprayed concrete lining.

Swelling ground conditions are typically related to marls, anhydrite, basalts and clay minerals such as corrensite and montmorillonite. Volume losses can reach more than 13% for corrensite

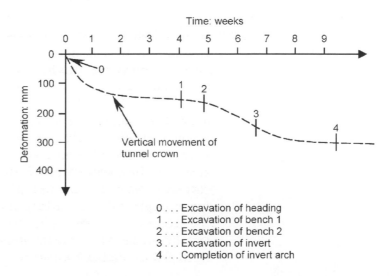

Fig. 5.6 Deformation over time from the excavation of the Arlberg Tunnel (after John, 1978)

(Wittke, 1990), although for basalts and marls the volume loss is much lower. Swelling is a stress-dependent process and calls for construction methods that limit the exposure of the excavation surface to water, for example, a sprayed concrete sealing layer. Water should be removed from the rock mass as quickly as possible. Where swelling is unavoidable, the linings should be designed specifically for the stress-dependent portion of volume change. Basic stress/strain relationships can be obtained from the Huder–Amberg (1970) test where the actual loading/unloading conditions are reproduced as accurately as possible.

5.3.2.3 Consolidation Time-dependent behaviour is also recorded in clays and a typical increase of 20–30% in pressure on segmental linings has been recorded over a period of about 20 years in London Clay (Barratt *et al.*, 1994). Most of this effect is related to consolidation and therefore there is considerable benefit in providing a full waterproof membrane or watertight gasket system in segmentally lined tunnels to prevent drainage, and so avoid attracting additional loads.

5.3.3 Groundwater

The presence of groundwater and associated seepage pressures often adversely affects the stability of tunnel excavations or linings (see Section 3.5.3 and Section 5.2 for effects during construction and methods of combating them). In such cases, TBMs capable of supporting the face, either earth-pressure-balance or pressurised slurry TBMs, or ground treatment in advance of tunnelling will be required.

Lining designs often have to consider the long-term hydrostatic pressures. In deep tunnels, for example hydroelectric projects where cover can exceed 1000 m, it is impractical to design linings for the full water pressure and pressure relief holes are required to avoid overstressing. In shallower tunnels, if seepage through the lining is unacceptable, a partial or full waterproofing system will be required. Full membranes are being used increasingly to control water inflows in new road and railway tunnels in countries such as Germany and Switzerland. This is often driven by environmental concerns (see Haack, 1991) and, at 2004, the maximum hydrostatic pressures designed for have been about 700 kPa. Beyond that, the lining designs become prohibitively expensive and at higher pressures it is difficult to construct membranes that do not leak (see Section 4.7).

5.3.4 Ground improvement and pre-support

If stand-up time is minimal and/or material strengths are low, ground improvement techniques or pre-support ahead of the face may be needed.

5.3.4.1 Forepoling and spiling Pre-supports in the forms of forepoling and spiling have been widely used in tunnelling over many years (Proctor and White, 1946). Forepoling is the installation of steel bars or sheets to provide crown support in soft rocks and stiff soils. Spiling is the insertion of elements of tensile strength, for example grouted GRP (glass reinforced plastic) dowels, into the ground, and is becoming more common, for example in Italy as part of the ADECO-RS (Analysis of Controlled Deformation in Rocks and Soils) full-face tunnelling approach (Lunardi, 2000).

Fig. 5.7 Low cover portals formed using canopy tubes – Crafers Highway Tunnels, Adelaide, Australia

Under the high stress, low rock mass strength environments, face dowels have been shown to be very effective in providing sufficient stability to allow full-face excavation. Glass-fibre dowels have the advantage over steel dowels of being easier to cut during excavation.

5.3.4.2 Pipe-roof umbrellas/canopy The use of canopy tube umbrellas has become common. The umbrella, which consists of closely spaced, grouted, steel tubes, is effective in controlling deformations and volume losses for a wide range of ground conditions by reducing dilation, improving face stability and increasing stand-up time. This is often needed at portals due to the low cover (see Fig. 5.7 for example).

5.3.4.3 Slot-cutting methods Slot-cutting on the perimeter of the arch, with filling of the slot with concrete, provides a continuous load-bearing arch that pre-supports the ground. It is applied to weak relatively homogeneous materials having sufficient stand-up time to allow the slot to stay open, for example the Perforex system at the Ramsgate Port Access Tunnel, UK.

5.3.4.4 Jet grouting In a similar manner to the pipe arch and slot cutting, jet grouting can be used from within a tunnel to form a strengthened arch in advance of excavation, particularly in very weak materials such as silts and sands. The grouting can also be done from the surface.

5.3.4.5 Face sealing A sealing layer of sprayed concrete or structural polymer membranes can be applied directly to the ground to prevent the failure of individual blocks. The former has been used successfully in soft ground and rock (Powell *et al.*, 1997) while the membranes have been used experimentally in mines to control rock burst.

5.3.5 Effects of ground improvement or water management on linings
The use of ground improvement techniques or groundwater control methods can have effects on the design of the shaft or tunnel lining. These may occur during the construction phase or over an extended period after completion of the shaft or tunnel. Each technique can affect the loading on the lining or have direct effects on the lining construction. The following effects represent the main areas to be considered during the design process.

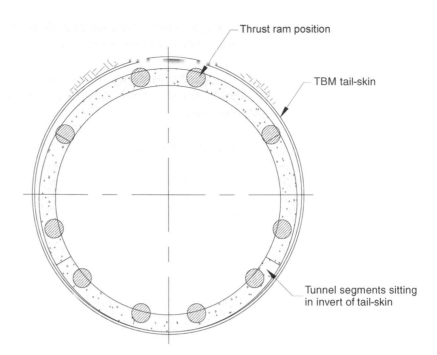

Fig. 5.8 Thrust ram eccentricity due to gravity

seals to prevent ground and water ingress tend to reduce this eccentricity.

In tunnels below the water-table in stable strata the tunnel lining may float to the crown of the excavation resulting in an inverse situation of ram eccentricity.

Other direct loads can be applied to the tunnel lining from the towed TBM back-up train. These loads may be significant if the load distribution on the gantries is not even. Gantries mounted onto skids are particularly prone to causing damage to the circle joints between segments.

5.3.6.2 Machine tunnelling in rock When a TBM is used to excavate segmentally lined tunnels in rock the same problems as outlined in Section 5.3.6.1 may be evident. However, where a rock TBM employs a 'friction gripper ring' between the machine and the excavated profile to react against the thrust, the excavation and face support techniques have little effect on the design of the lining, because it is installed independent of the TBM.

5.3.6.3 Hand tunnelling Manual excavation is carried out in stable or treated ground with protection provided to the miners by an open-face shield or timbered support to the excavation. If a shield is used, progress is achieved by thrusting off the previously constructed segmental linings. The problems discussed in Section 5.3.6.1 should be considered.

In a timbered excavation the timbering may transfer loads onto the tunnel lining. The walings and soldiers of timbered face support, for instance as may be required at weekend stoppages, may transfer large longitudinal loads to the tunnel lining. While the effect of these loads is usually small compared to the permanent loads applied to the lining, this should be checked. The eccentricity of the loading and ground loading on the lining at that time should be considered.

Hand excavation is becoming less common in modern tunnelling due to potential problems with Hand Arm Vibration Syndrome (HAVS) from pneumatic manually operated tools. The designer should, wherever possible, specify a tunnel configuration where mechanical excavation and lining erection techniques can be employed.

5.3.6.4 Drill and blast It will be necessary to consider the effect of blast pressures and flying debris on the previously installed supports. Arch supports should be adequately braced and tied to prevent overturning during the blast. The effect of blasting on rock-bolts and anchorages must be considered. The use of non-destructive testing of critical anchors between blasting rounds may be necessary to prove that their integrity has not been compromised.

5.3.6.5 Sequential excavation Excavation in sequenced stages is often employed together with other measures, such as rock-bolting, mesh and lattice beams, in order to increase excavation stability where early support is required.

The detailed design of sprayed concrete linings is dependent upon the method of excavation and the stability of the ground. Each stage of tunnel excavation and lining construction must be considered for stability, and also its effect on ground movements. Early closure of the ring is critical to the control of ground movements in shallow tunnels. With the exception of small diameters, excavation will be in the form of a series of benches (see Fig. 5.9).

In some cases temporary drifts may be required to ensure face stability and reduce surface settlement. It is essential to consider the stability of the lining at all stages including partial or full removal of the temporary drift lining and connections to the subsequent lining. The size of the drifts should be designed to enable safe excavation and lining installation, given the equipment to be used.

In less stable conditions, the strength-gain characteristics of the sprayed concrete should be considered in relation to the tunnel advance rate and the rate of loading from the ground.

The installation of rockbolts, lattice girders and mesh requires men to work at the partially unsupported face. While safety is likely to be achieved with the application of an immediate sealing layer of shotcrete on the excavated face, there is still a risk small falls of materials could occur and injure persons in the vicinity. The likelihood and consequences of this risk will depend on a number of factors including the ground characteristics and sequences of work involved.

5.3.7 Choice of lining systems

As discussed in Chapters 2 and 4, the design life and use of the structure have a major impact on the choice of lining system. The lining system must be constructed in accordance with a specification that reflects best practice, for example the BTS *Specification for Tunnelling* (British Tunnelling Society, 2000). The final choice of lining system will be influenced by the expected ground conditions and cost, the Contractor's preference, if several support methods are available, and the choice of construction method.

5.3.7.1 Segmental linings Circular gasketted segments offer an economical and efficient method of constructing tunnel linings, especially in soils and weak rocks (see case histories in Chapter 10). Modern segmental linings are generally a robust solution and can be used to provide a 'one-pass' solution. They are well suited

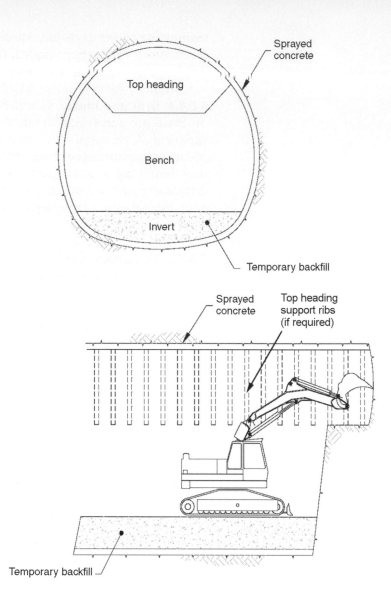

Fig. 5.9 Sprayed concrete lining in benches and temporary drifting

to mechanised tunnelling where long tunnel sections and fast progress are required. More importantly, they offer a safe construction method that does not involve placing personnel at an unsupported tunnel face.

Many hundreds of kilometres of segmentally lined tunnel have been constructed throughout the world using this technology. Two-pass linings consist of a concrete lining with a inner lining of brick or in situ concrete. One-pass linings of precast concrete or cast-iron (including SGI) rings have been used, for example for tunnels of the London Underground system. The segmental linings may be bolted or expanded. They may have one, two or no water-proofing gaskets, depending on watertightness criteria.

5.3.7.2 Sprayed concrete linings   In rock tunnels the rock itself is the principal support with supplementary support installed as required. Sprayed concrete is often used for the supplementary support and is very suitable for tunnels where the support required may vary. The time-dependent response of sprayed concrete has been found to work well in highly stressed rock (see Rabcewicz, 1969). Whether these are one-pass or two-pass systems depends on the design details of each individual case.

The use of sprayed concrete lining (SCL) systems is becoming increasingly common in soft ground because of the flexibility that

Tunnel lining design guide. Thomas Telford, London, 2004

SCL offers in terms of the shape of the tunnel and the combination of support measures. Sprayed concrete lining is particularly cost effective for short tunnels and junctions. Two-pass systems have traditionally been used in SCL tunnels where the ground is not largely self-supporting. The primary support systems are designed to maintain a stable excavation so that a permanent lining can be placed.

In soft ground one-pass sprayed concrete linings have rarely been used because of concerns over the durability of the lining. Traditionally the linings contain lattice girders for shape control and mesh for reinforcement. Voids behind the bars (shadowing) raise the possibility of corrosion. This can be avoided by using steel-fibre reinforced sprayed concrete linings with no lattice girders. However, there are a number of technical issues with this type of construction including the key issues of control of shape and thickness and quality of construction.

5.3.7.3 Cast in situ linings Cast in situ concrete linings usually require temporary support to secure the excavation, and so are the permanent part of many two-pass systems (see Section 5.6).

5.4 Segmental linings

Segmental linings are the commonest form of lining for soft ground tunnels, particularly for relatively long lengths where the economics of using a TBM are most advantageous. The design of a segmental ring not only requires a structural analysis for the ground loads and the TBM ram loads applied to the segments, it also requires the designer to consider the total process of manufacture, storage, delivery, handling and erection as well as the stresses generated by sealing systems and bolts or other erection aids.

5.4.1 Transport, handling and erection

Transport, handling and erection are critical operations and can often be the determining factors of the design, particularly for a precast concrete segmented lining. These operations should not be regarded as a detail to be dealt with later, but should be considered early in the lining design process.

5.4.1.1 Lifting and handling at the factory It is often necessary, for reasons of economy, to demould and handle precast concrete segments at an age of approximately 18 hours after casting. In order to make this possible without causing damage to the segments care must be taken in specifying a suitable concrete mix, section aspect ratio (length/thickness) and reinforcement arrangement. Suitable lifting methods and lifting points need to be identified and designed as part of this process, (see Fig. 5.10).

5.4.1.2 Transportation and storage above ground Suitable orientations and permissible stacking heights must be specified to suit the concrete strengths at the time that transport to and storage on site will take place, (see Fig. 5.11). Impact due to stacking and during transport should be considered and suitable spacers specified to prevent damage during these operations (see Fig. 5.12).

5.4.1.3 Handling in shafts and tunnels Safe handling points should be identified and designed to resist the loads imposed from the handling and transport system both above and below ground. Segments should be sized to allow manoeuvring within

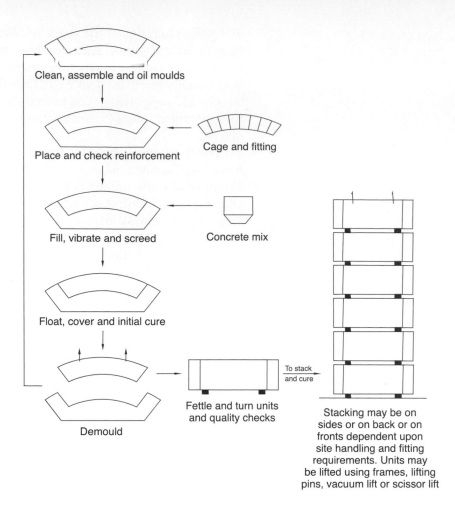

Fig. 5.10 Typical precast concrete segment production process

the confined spaces of a tunnel and tunnel back-up system while maintaining safe and free movement of personnel, materials and equipment.

5.4.1.4 Shield and hand erection Buildability and shield compatibility are essential elements of any lining design. Segments must be designed to be easily and safely located and fixed in position using cast-in components or other suitable means that are compatible with the method of erection.

For mechanised shield erection the provision of a central lifting point will be necessary, or safety dowels must be provided for use with vacuum erectors. The cast-in lifting point must be designed

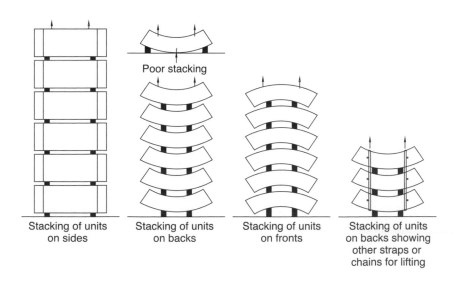

Fig. 5.11 Segment stacking variations and damage due to poor stacking

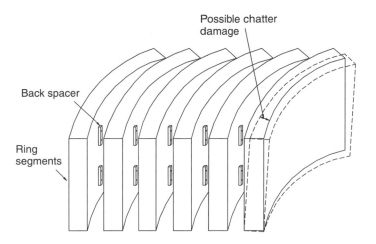

Fig. 5.12 Stacking concrete segments with back spacers to prevent chatter damage during transport

for pull out, moment and shear force induced by the segment self-weight and the impact loading associated with the erector use. Accelerations from the hydraulic manoeuvring rams on the erector are likely to result in loads greater than the segment dead weight; therefore a factor of safety of at least 3.0 on the dead loads is usually required.

5.4.1.5 Joints Design of the joints should provide for fast and durable connections with sufficient strength to meet erection sequence support requirements and to maintain compression of the sealing gaskets. Particular attention must be paid to the design of longitudinal joints. High level contact stresses due to joint geometry and ring build may cause circumferential cracking due to high tensile stresses. Pads can be used to reduce these stresses. Gasket compression has an important influence on the joint design, as it requires large forces to close joints and then hold them together. Positioning and size of gaskets for sealing can significantly reduce the cross-sectional areas of joints available for the transfer of compression loads. Relief of loading of the areas at the extrados of the segment behind the gasket can help reduce damage caused by gasket compression. Hence the joint connection, strength, number and position must be designed to ensure and maintain adequate gasket compression.

Consideration should also be given to the relief of loading at the edges of segments to minimise spalling when ram loads are applied (see also Section 5.3.6.1). When completing the ring erection, key sizes and angles must be compatible with the available tail-skin space and shield ram-travel when a ram is used to place the final unit.

Provision for bursting steel may be necessary for large ram loads and loading pads can be helpful in reducing segment damage.

5.4.1.6 Fixings To achieve continual production, tunnel services, for example lighting, power, telephone, compressed air, ventilation, need to be constantly extended to keep them close to the working face. As the tunnelling method becomes more sophisticated with the introduction of pressurised TBMs so the number of services increase. For safety and maintenance reasons these services are normally supported by purpose-designed brackets bolted to the tunnel lining particularly in tunnels lined with one-pass linings. (Care should be taken to ensure that fixings used to support services do not have a detrimental effect on the lining.)

In a reinforced concrete lining, whether in situ or precast, the intrusion of a corrosive fixing could, over time, reduce the structural integrity of that lining. Therefore, wherever risk of corrosion arises, stainless-steel fixings or removable fixings should be used. Attention should still be paid to the potential of mechanical fixings to effectively reduce the cover over reinforcement.

To achieve this precast concrete manufacturers can provide preformed holes for fixing bolts or small indentations on the inner face of a tunnel lining segment to mark areas where drilling will not encounter reinforcement and hence will not cause structural instability.

5.4.1.7 Sealing gaskets Ethylene Polythene Diene Monomer (EPDM) compression gaskets or hydrophilic seals are commonly specified so as to provide waterproof joints between adjacent segments in a precast concrete tunnel lining. The performance of these seals with respect to water pressure, gasket compression characteristics and joint gap tolerances is an important part of the lining design. The long-term durability and deterioration of the performance of the seal due to creep and stress-relief should be taken into account. The likely fluctuation in water level will dictate the type of gasket to be employed. Hydrophilic seals may deteriorate if repeatedly wetted and dried. Different hydrophilic seals may be required for saline and for fresh water as the performance of hydrophilic seals can be affected by the salinity of the groundwater.

5.4.1.8 Tolerances Particular attention must be paid to dimensional tolerances required by different TBMs, especially those containing tail-skin brush-seals (to prevent loss of seal grease on rough-surface segments), and to segment lengths in tight back-up systems (see BTS *Specification for Tunnelling* (British Tunnelling Society, 2000)).

5.4.2 Annulus grouting of segmental tunnels

5.4.2.1 Design requirement The purpose of grouting is to fill the annulus around the tunnel lining and achieve the following:

- *maintain the tunnel ring shape*
- *distribute ground pressures evenly onto the lining while maintaining stability*
- *reduce seepage/ground water inflow*
- *limit surface settlement (see Section 5.3.5.1).*

It is necessary to consider the proposed annulus grouting system and method during the detailed design of the tunnel lining.

5.4.2.2 Method of grout placement
Through the TBM tail-skin In this method grout is injected through grout lines integral with the tail-skin; this limits the amount of unsupported ground. This technique can be used on larger machines in areas where the control of surface settlement is critical or where, because of their size, the risk of excessive surface settlement is acceptable. The use of through the tail grouting on smaller machines, circa 3 metres diameter, can often introduce additional problems as the TBM needs to be bigger to enable grout pipes of sufficient size to be introduced into the tail. By increasing the TBM O.D. the thickness of the annulus is increased which, if not filled properly, could give rise to increased settlement.

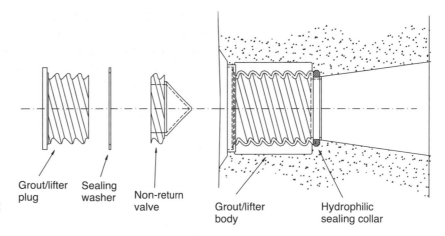

Fig. 5.13 Typical cast-in fitting for grouting

Grout/lifter plug Sealing washer Non-return valve Grout/lifter body Hydrophilic sealing collar

Problems may arise, when using brush seals at the rear of the TBM, if grout is allowed to set in the seals. The contamination of the brush seals with solidified grout will severely affect the effectiveness of the seals and may result in an excessive consumption of seal grease.

Through the tunnel lining Grout can be injected under pressure through grout holes cast in the tunnel linings (see Fig. 5.13). This method of grout placement is usually employed in relatively stable ground or where the control of settlement is less critical.

5.4.2.3 Geometry of grouting array The positioning of grout holes within tunnel rings should be considered at an early stage. Grout sockets are often cast into segments and utilised as lifting/ handling sockets. However, it may be preferable to locate the grout points closer to the trailing edge of the rings since this facilitates early placement of the grout as the completed tunnel ring leaves the tail-skin. Also the orientation of the grout injection points should be checked against the back-up equipment to ensure that all points are accessible.

The grout arrays at junctions with shafts should also be carefully considered to ensure that an adequate seal can be attained. It is common for segments containing a number of grout injection points to be specified in these areas.

5.4.2.4 Grouting pressures The grouting pressure must be greater than the external hydrostatic pressure acting on the lining at the time of grouting. In larger diameter tunnels the difference in hydrostatic pressure between the crown and the invert can be considerable. However, it must be noted that excessive pressure and/or injection of too much grout may cause damage to the lining or cause ground heave, which could affect surface or sub-surface structures and services.

5.5 Sprayed concrete linings

In his early work in rock tunnels, Rabcewicz (1969) recognised that sprayed concrete was a material well suited to tunnelling for the following reasons.

- **Permanent lining** Sprayed concrete is a structural material that can be used as a permanent lining.
- **Early age strain** The material behaviour of sprayed concrete, which is initially soft and creeps under load but can withstand large strains at an early age, is compatible with the goal of a lining which permits some ground deformation and therefore stress redistribution in the ground.

- **Deformation** The material behaviour, specifically the increase in stiffness and strength with age, is also compatible with the need to control this deformation so that strain softening in the ground does not lead to failure.

Sprayed concrete linings can be formed as and when required, and in whatever shape is required. Hence the geometry of the tunnel and timing of placement of the lining can be tailored to suit a wide range of ground conditions. Sprayed concrete can also be combined with other forms of support such as rock-bolts and steel arches. Also:

- *there are lower mobilisation times and costs for major plant items*
- *the same equipment can be used for shaft construction as well as tunnelling*
- *the method is compatible with the Observational Method, CIRIA (1999) which permits optimisation of support (and therefore costs) during construction*
- *the freedom of form permits tunnels of varying cross-sections and sizes and junctions to be built more quickly and cost-effectively than if segmental or cast in situ linings are used.*

5.5.1 Potential weaknesses

Following the collapse of a series of soft-ground SCL tunnels in 1994, this construction method came under intense scrutiny. Reports by the Health and Safety Executive (1996) and the Institution of Civil Engineers (1996) have established that SCL tunnels can be constructed safely in such conditions and the reports provided guidance on how to ensure this during design and construction. The reports drew attention to weaknesses of the method.

- **Operator** The person spraying the concrete, the 'nozzleman', has a considerable influence over the quality of the lining so the method is very vulnerable to poor workmanship. This is particularly true for certain geometries where the geometry makes it difficult to spray the lining or to form clean joints.
- **Need for monitoring** The performance of the linings and ground must be monitored during construction to verify that both are behaving as envisaged in the design. The data from this monitoring must be reviewed regularly within a robust process of construction management that ensures that abnormal behaviour is identified and adequate countermeasures are taken.
- **Difficulty of instrumentation** It is difficult to install instrumentation in sprayed concrete linings and to interpret the results (e.g. Clayton *et al.* (2000)).
- **Material behaviour** It is difficult to design SCL tunnels because of the complex material behaviour of sprayed concrete.

Other disadvantages of this method, specifically as applied to soft ground, are as follows.

- **Closure of the ring** It is of critical importance to minimise deformations, otherwise strain-softening and plastic yielding in the ground can lead rapidly to collapse. Complex excavation sequences can lead to a delay in closing the invert of the tunnel, and forming a closed ring (see also Section 5.3.6.5). This delay can permit excessive deformations to occur.
- **Control in shallow tunnels** In shallow tunnels the time between the onset of failure and total collapse of a tunnel can be very short. Consequently, much tighter control is required during construction.

Advance rates are generally slower than for shield-driven tunnels so SCL is not economic for long tunnels (i.e. greater than about 500 m) with a constant circular cross-section unless the ground conditions preclude the use of a TBM. A higher level of testing during construction is required for quality control compared to segmental tunnel construction. Mix design tests are needed before the works actually commence. The surface finish of a sprayed concrete lining is quite rough and may contain protruding steel fibres and so be unsuitable for a final finish unless a smoothing layer is applied.

Following the HSE and ICE reports, a considerable amount of guidance has been produced on such subjects as certification of nozzlemen (see Austin *et al.*, 2000 for the latest guidance), instrumentation, monitoring (Health and Safety Executive, 1996) and risk management.

5.5.2 Design issues

5.5.2.1 New Austrian Tunnelling Method (NATM)
The NATM principles established by Rabcewicz (1969), Pacher (1977) and Müller (1978), and reiterated later by Golser (1996), acknowledge that there is a relationship between the applied stresses, the ground properties and the required capacity of the support system. Developing a system that incorporated all of these elements into an economical design was achieved through the use of sequential excavation methods, support systems and lining geometries that controlled deformations. The ground will inevitably deform and there is a re-distribution of the in situ ground stresses. Managing this and creating a self-supporting stable arch are the cornerstones of the NATM. This is a complex three-dimensional and time-related process. The critical element is to control the strains progressively around the opening so that the load-bearing capacity of the ground is preserved to the greatest degree possible. If this is achieved, a cost-effective excavation and support system will result.

5.5.2.2 Sprayed concrete lined tunnels in soft ground
Sprayed concrete lined tunnelling is only possible where the stand-up time is sufficient to allow a limited unsupported advance. In terms of soft ground, this usually limits the technique to stiff clays, some dense sands and gravels, and weak rocks, unless sufficient ground improvement has been achieved. Sprayed concrete lined tunnelling has proved to be an effective method of construction under these conditions. Some arching will occur in the ground and the design should seek to maximise the proportion of the ground loads that are carried by the ground. The subject of ground/support interaction is discussed in more detail in Section 5.3.1.

5.5.2.3 High-stress situations
Under high confinement conditions, for example in the Alps, the stress readjustments may take months to stabilise and reach equilibrium. Deformations imposed on any support system would normally exceed elastic limits. Designing for large deformations requires a clear understanding of the methods employed for controlling deformations and how the load-transfer process can be used. Strain softening is likely to occur. Providing that the ground behaves almost as a continuum, the load-bearing capacity of the zone of residual strength that develops can be used. The principal requirement is to maintain

Fig. 5.14 Subdivision of a SCL tunnel face to maintain stability

the shear strength of the ground within this zone at an appropriate level using a flexible support system.

5.5.2.4 Construction sequence The construction sequence is a major influence on the loads on a sprayed concrete lining. In soft ground, early closure of the ring (or a part of the tunnel, such as a side gallery) is critical to the control of ground deformations. This also affects the short-term loads on the lining (see also Section 5.3.6.5).

5.5.3 Detailing

Sprayed concrete lined tunnels often have subdivision of the face (see Fig. 5.14). These subdivisions should be sized according to the face stability of the ground and the size of the construction equipment.

Ideally a sprayed concrete lining functions as a shell structure. This is why it is so well suited for the tunnel junctions. To achieve this there must be structural continuity across the many joints. Starter bars are used to achieve the required laps with steel reinforcement where applicable. It is important to keep the design of the joints as simple as possible in order to avoid construction defects (such as shadowing and trapped rebound). Where possible joints should not be placed in highly stressed parts of the lining.

Unlike segmental linings, there is a great potential for variability in the shape of SCL tunnels. Poor shape control can lead to stress concentrations. Therefore lattice girders or steel arches are often included to ensure that the correct shape of the lining is formed. These components are not normally included in structural calculations.

5.5.4 Performance requirements

Because of the uncertainties that surround the design of SCL tunnels, instrumentation is installed to verify that the tunnel is performing as intended. The monitoring data should be reviewed on a daily basis against the Key Performance Indicators (see Section 5.8). Trends in monitoring data are as important as the absolute values of the data. Further details of the management of SCL construction can be found in *Safety of New Austrian Tunnelling Method (NATM) tunnels*; Health and Safety Executive

(1996) and in *Sprayed Concrete Linings (NATM) for Tunnels in Soft Ground*, Institution of Civil Engineers (1996).

5.6 Cast in situ linings

In the past in situ linings were formed of brick or stone masonry. Now they are of unreinforced or reinforced cast concrete. The lining is constructed within some form of (temporary) ground support that has been installed to create a safe working environment.

5.6.1 Design requirements

The design of in situ linings is relatively straightforward if the initial ground support is assumed to be temporary only. The in situ lining must be designed to carry all the loads for the full design life of the tunnel. Normal design codes apply for design detailing since the lining does not carry external loads until the concrete has cured. Temporary loadings on the lining and its formwork during casting can be dealt with using standard design methods for above-ground structures.

The situation is more complicated if the initial ground support is assumed to carry part of the long-term loads. The initial support and the in situ lining then act as a composite structure. The nature of the load sharing between the two linings will require careful consideration since it will depend on the specific details of each case.

5.6.2 Grouting

Grouting is required to ensure that the in situ lining is in full contact with the initial ground support and ground. Bleed pipes will be required to ensure that air does not become trapped during grouting or concreting. The grouting pressure should be limited so that it does not damage the new lining, and the grouting arrangement should be compatible with the waterproofing design.

A simple relationship has been proposed that provides an estimate of the eccentricity of thrust in relation to the size of void left behind the lining due to incomplete grouting (Bickel *et al.*, 2002)

$$e = C^2/8R$$

where e is the eccentricity in metres, C is the chord length in metres and R is the tunnel radius in metres.

5.7 Special constructions

5.7.1 Shafts

The method of excavation and support adopted during shaft construction has a direct influence on the design of the shaft lining. In the permanent condition shafts are usually subject to uniform radial ground and groundwater loading, but in the temporary condition the imposed loads due to the construction operations may create a more onerous loading regime.

The designer of the shaft lining should be familiar with the different construction methods available and how they affect the design of the shaft lining. The possible methods of constructing the tunnel entry/exit and the effect on the design of the tunnel lining should also be considered at an early stage to avoid costly delays at a later date.

5.7.1.1 Segmental shafts by underpinning
In competent ground that can stand unsupported for the required depth and time to allow excavation and erection of a number of lining segments,

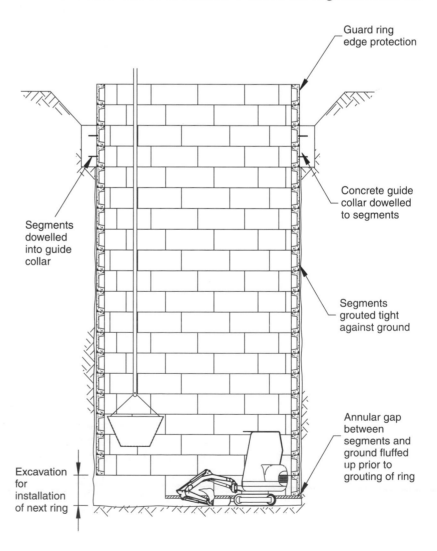

Fig. 5.15 Underpinning hook for use with standard bolted shaft segments (left) and underpinning device for use with smooth-bore segments (right)

Bolthole

Counterbalance force

Threaded grout socket

Counterbalance force

underpinning methods can be employed. Segments are lifted individually into place using bespoke underpinning hooks and counter balance arms if required, using a crane (Fig. 5.15). The effects of handling in this way should be considered.

It is necessary to fill the annulus behind the underpinned segments with grout to transfer a uniform ground loading and reduce surface settlements. In practice this takes place at the end of the working shift, or otherwise prior to the onset of instability of the vertical excavation surface.

There may be a number of rings effectively suspended from previously grouted rings or the guide collar (Fig. 5.16). The rings should be considered as being suspended until such time as sufficient friction can be mobilised between the segments and the

Guard ring edge protection

Concrete guide collar dowelled to segments

Segments dowelled into guide collar

Segments grouted tight against ground

Annular gap between segments and ground fluffed up prior to grouting of ring

Excavation for installation of next ring

Fig. 5.16 Typical shaft installed by underpinning techniques

Tunnel lining design guide. Thomas Telford, London, 2004

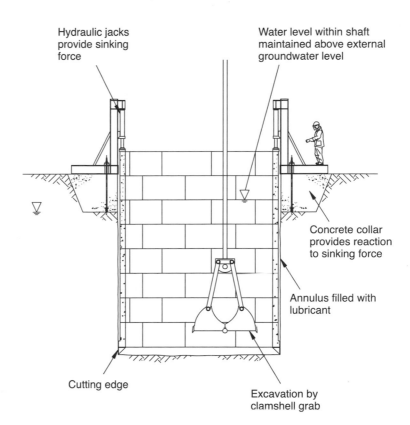

Hydraulic jacks provide sinking force

Water level within shaft maintained above external groundwater level

Concrete collar provides reaction to sinking force

Annulus filled with lubricant

Cutting edge

Excavation by clamshell grab

Fig. 5.17 Typical shaft installed by (wet) caisson method

ground, that is upon setting of the grout. The effect of this dead load on the segments and their connections must be considered.

5.7.1.2 Segmental shafts by caisson sinking Caisson-sinking methods are generally employed in those ground conditions where a vertical excavated face is difficult to achieve, or where base stability is of concern due to water pressure.

In water-bearing strata where dewatering within the shaft would cause basal instability, 'wet caisson' techniques are appropriate, that is a positive head of water is maintained (Fig. 5.17). In water-bearing strata that can be dewatered without causing basal instability, but where side-wall instability is a problem, 'dry' caisson methods can be employed.

In both cases a segmental lining is erected at the surface and sunk into position using kentledge or hydraulic jacks to assist the self-weight of the shaft to overcome ground friction. Excavation under water is usually carried out from the surface by the use of a crane-mounted grab. Air lifting pumps or a long-reach excavator can also be employed. If the shaft is maintained dry, excavation may be by the use of a mechanical excavator within the shaft and removal of muck by skips or a vertical conveyor.

If hydraulic jacks are used, the effect of the application of vertical load to the caisson at discrete locations must be considered. Similarly the application of kentledge load through a frame should be considered. The determination of the magnitude of these loads is problematical and is usually based on previous experience.

The verticality of the shaft has an influence on the structural integrity of the shaft. The application of a vertical shaft-sinking load to an out-of-plumb shaft may cause an onerous loading condition. The combined caisson-sinking loads, and earth pressures should be checked, assuming the shaft is out of the vertical by 1 in

300, and a diametric tolerance of 1% of the diameter or 50 mm, whichever is the greater.

If the caisson cutting edge encounters an obstruction, for example a boulder or rock intrusion, caisson sinking may cease. It will be prudent during the design of the shaft lining to check that the connections between segments are adequate to transfer the loads should this occur and the caisson is supported on one side only.

5.7.1.3 Other methods of shaft sinking There are several other methods available for the construction of shafts including diaphragm walls, bored piled walls, steel sheet-pile techniques, shaft drilling with casing and, in self-supporting rock, raise-boring. The design and construction of soft-ground piled shafts is similar to the methods employed in the design and construction of earth-retaining walls and is outside the scope of this Guide.

Particular attention should be paid to construction tolerances, particularly for shafts constructed with limited space for the construction of walings, or diaphragm-wall shafts dependent on a continuous ring of concrete in compression. Typical achievable construction tolerances are given in the Institution of Civil Engineers (UK) *Specification for Piles and Embedded Retaining Walls* (Institution of Civil Engineers, 1996b) and *Specification for Diaphragm Walling*, and the Federation of Piling Specialists *Guidance Notes*.

5.7.2 Junctions and portals

The construction of a junction between the tunnel and another tunnel, a shaft, or other underground structure is often the critical operation in tunnel construction, in terms of the stability of the ground and the structure. All too often the importance of this operation is not fully understood at the design stage resulting in excessive costs and delays.

The designer must consider not only how the integrity of the structure is to be maintained in both the short and long term, but also how the stability of the ground is to be maintained to prevent ingress and associated surface settlement or collapse.

Due consideration must be given to the following.

5.7.2.1 Structural stability How are the temporary and permanent loads in the structure to be transferred around the tunnel eye or junction without causing distress to either the structure or the tunnel?

Whenever an opening is made in an existing underground structure the loads already developed in the structure must be considered and transferred accordingly. The loads that have been developed will be dependent upon the ground conditions, the groundwater regime and, in some cases, the period for which the structure has existed and the period that has lapsed since the opening was formed. In addition the permanent design loads in the structure must be transferred by the opening support structure.

Determination of the loads that have developed in a structure is problematical, particularly in cohesive strata. In structures with high groundwater levels, however, the contribution from the ground pressures will be negligible compared to the hydrostatic head of water. Provided that a suitable load factor is applied to the water pressure a robust design will follow. It can be seen therefore that in a cohesive material there will be a clear difference between the temporary and permanent loads whereas, in a granular material below the watertable, the distinction is less obvious.

The permanent works in a structure in granular material below the watertable must be in place prior to the formulation of the opening, unlike in cohesive strata where temporary works can be utilised to support the opening until such time as the permanent structure is in place.

5.7.2.2 Ground stability
How are ground deformations and ground loss due to groundwater ingress to be controlled?

The measures to be taken to prevent instability of the surrounding ground will depend upon the type of ground at the opening horizon, the type of ground both immediately above and below the opening, the stand-up time of the face, and the groundwater regime.

In unstable water-bearing strata it will be necessary to either treat the ground outside the structure, by grouting, dewatering or freezing techniques, or provide full support to the face at all times during the break out.

Even when breaking out from a structure in a stable strata, such as London Clay, care must be taken because there will inevitably be a zone of disturbed ground around the structure. A very thorough grouting exercise during the construction of the structure may not prevent water being drawn down from an aquifer above the structure, with consequent effects.

In water-bearing ground a suitable system of seals must be provided to exclude groundwater.

5.7.2.3 Constructability
How are the resultant measures to be incorporated into a robust, safe and economical design?

Consideration must be given to constructability issues, particularly in an environment that may have limited room and where normal cranage is not available to aid in the construction process. When designing an opening support structure in a shaft, surface cranage is likely to be available, and hence the size of sections and other elements is of less concern.

When designing an opening support structure in a tunnel, however, cranage is not likely to be available, and steel sections, for example, must be sized to enable installations using winches and pulleys.

Thought must be given when detailing in situ concrete to ensure that voids are not left unfilled. It is usual to specify sloping faces to encourage air to be displaced and not trapped.

5.7.3 Portals, launch chambers and reception chambers
Portals present particular challenges because of the low cover and potentially poor nature of the ground, for example due to weathering. The lining usually must be designed to carry the entire weight of its overburden since the ground may not be able to arch over the tunnel. Additional measures such as canopy tubes may be required to guard against instability in the crown of the tunnel.

For TBM tunnels special measures may be required at the launch and reception points of each drive, either because of the low cover and inferior ground conditions or because of the need to maintain the pressure around a closed-face TBM. Where expanded segmental linings are used for the main tunnel, it is common practice to used bolted segmental rings within a distance of about two tunnel diameters of the portals, to safeguard against differential movements. However, in general the design of the lining is not affected. Therefore these measures are not discussed further.

5.7.4 Tunnels in close proximity

If two tunnels are constructed within two diameters of each other (measured from centreline to centreline) both tunnels can be expected to be affected by the presence of the other one. The load on both tunnels will be increased. The first tunnel will tend to move towards the second tunnel due to stress relief and it will suffer additional distortions. The magnitudes of these effects can be estimated using simple analytical tools (Hoek and Brown, 1980) and the principle of superposition or the situation can be modelled explicitly using numerical models (Soliman *et al.*, 1993).

The construction of a tunnel will also lead to disturbance of the ground around it. When planning the construction of multiple tunnels, a sequence of construction that avoids tunnelling through highly disturbed ground should be sought. For example, if there are three parallel tunnels, it is preferable to build the central tunnel first, rather than last.

5.7.5 Jacking pipes

Pipe jacking is a specialised installation technique that has been carried out for over 40 years but it was only the subject of detailed research and development in the UK over the last 15 years. Early research at Oxford University resulted in significant advances in understanding the build up of stresses and strains within pipes (Ripley, 1989) and further work was carried out on the effects of different types of ground on the magnitude of applied loads on the pipe by Norris (1993). General guidance on design for this specialised area of underground construction is given in the Pipe Jacking Association's *Guide to Best Practice for the Installation of Pipe Jacks and Microtunnels* (Pipe Jacking Association, 1995).

5.7.6 Pressure tunnels

This aspect of tunnel design attracted attention in the 1960s and 1970s because of numerous incidents of leakage and penstock failure on hydroelectric projects. An in-depth study of the criteria applied to the design of pressure tunnels is found in Brekke and Ripley (1987).

5.7.6.1 Rock mass confinement   Of most importance in design is the rock mass confinement. In this context, confinement is defined as the ability of the rock mass to withstand the internal pressures generated by a static water level combined with the surge pressure generated when the main inlet valves of pressurised conduits are closed. Currently projects with pressure heads of up to 1000 m are in operation.

A number of criteria have been developed to guide selection of the vertical cover for unlined pressure tunnels so as to avoid problems with leakage. The one generally accepted is the Norwegian criterion (Broch, 1984)

$$C_{rm} = (h_s \gamma_w F)/\gamma_r \cos \beta$$

where

C_{rm} is the minimum rock cover in metres
h_s is the static head in metres
γ_w is the unit weight of water (kN/m^3)

Tunnel lining design guide. Thomas Telford, London, 2004

γ_r is the unit weight of rock (kN/m^3)
β is the slope angle (degrees)
F is the factor of safety.

This criterion must be used with care, and an additional check should be undertaken to make sure that the unfactored minimum principal stress (σ_3) along the tunnel alignment is never less than 1.3 times the maximum internal hydrostatic pressure. This check should consider the lateral as well as vertical cover and make sure that there are no interconnected major discontinuities or any significant deformable zones intersecting the tunnel. Most deformable zones will require suitable support measures to prevent deterioration/erosion of the rock mass.

5.7.6.2 Impermeable lining Where the confinement is not adequate, and to prevent leakage into the rock mass, an impermeable lining system is necessary (Brekke and Ripley, 1987). This usually requires either concrete or concrete encased steel linings. The design of such systems depends on the confining pressures and tensile strength of the rock mass and the extent to which the lining and surrounding medium distribute the internal pressures.

Typically the measured load transfer from steel to concrete is 50–90% of the internal pressure. The safe assumption is that the steel has to carry all the external water and ground pressure (following emptying of the tunnel for maintenance), and must also sustain handling and erection loads. If the ground conditions are good, and load transfer is accepted, most designers limit load sharing to 50%. However, this assumption should be carefully checked to assess the ability of the concrete (and any disturbed rock) to carry tension, the impact of any annular gap that forms at the concrete/steel interface, and the effect of external groundwater pressures. More sophisticated methods, of prestressed concrete linings that can control leakage, have been used successfully and a full discussion of this approach is given by Seeber (1985a; 1985b). The decision on the lining method will be a balance between cost and technical factors and the risk involved.

Occasionally, thin steel linings are installed to control only leakage. In this case buckling of the steel lining could occur if unbalanced external pressures are applied and should be checked.

Often designers assume that a rock mass, particularly for shafts, can be grouted to contain leakage if it occurs during operation. It is important to stress that this is not usually successful and can become an extremely expensive activity that may still require a steel lining to remedy the problem. The potential disruption to operations and loss of revenue need to be fully evaluated to put such risks into context.

5.8 Design guidelines on performance requirements

The preceding sections and Chapter 6 discuss the issues that need to be taken into consideration and the methods of analysis that can be used to support the tunnel lining design. All of the methods have to be applied in accordance with accepted standards and codes of practice to provide the designer and client with the reassurance that the lining design is robust.

The term 'robust' is not easy to define. Within the context of fundamental ground/support interaction principles, there has to be reasonable certainty that the support system can cope with potential failure mechanisms and adjust to the normal variations associated with the method of construction. An adequate factor of safety is achieved using appropriate standards and codes of practice on the basis that they reflect experience and best practice. It is possible that standards and codes of practice cannot cover all eventualities, so factors of safety will vary depending on the failure mechanism. Judgement is necessary in these circumstances and when that is coupled with good risk management it should be possible to develop a robust design that is not over-conservative.

Some aspects of a design can be difficult to quantify, and local practices and materials may affect the issue of robustness. Quality control is also important and many residual risks can be eliminated through the use of management systems that build in checks and procedures targeted at those risks. Safety-related problems sometimes involve human error and a robust design will include a consideration of this factor.

5.8.1 Key Performance Indicators

Limits on Key Performance Indicators (KPIs) are commonly used when specifying the required performance for a tunnel. This was an important feature of NATM constructions on the London Heathrow Express and Jubilee Line Extension. In-tunnel deformations are generally the most reliable indicator of potential problems, specifically from trends but also recognising that absolute values and critical strains should be taken into consideration. However, criteria are also often required for surface settlement in urban areas (see Chapter 7).

The following sections look briefly at performance-related criteria. This knowledge enables the designer to specify limits on deformations or distortions in order to ensure a controlled response during construction. This is the final element of a 'robust' design.

5.8.2 Ground response

As discussed earlier, the performance of the ground will depend on several factors such as the geomechanical properties of the materials (including the strength–stress ratio) and the timing of installation of support. These will provide an indication of whether a plastic zone could form and, if so, its likely extent considering the benefit of radial confinement from the installed support. In most cases a plastic zone is beneficial in re-distributing stresses providing that the yielding is controlled.

In many rock tunnels there will be an elastic response and therefore the strains in the ground around the tunnel will be nominal. Stability under these conditions is related only to the ability of individual blocks or wedges to slide or fall, that is they are kinematically admissible. The failure of an individual block is not of itself critical in terms of tunnel stability unless it is a key block that leads to a loss of arching action.

In general, designers are required to assess the safety risk in tunnels and therefore it is important to look at performance requirements and specify limits on deformations or distortions in order to ensure a controlled response during construction. These aspects are discussed in the following sections.

5.8.3 Lining flexibility

A general discussion of the effect of lining and ground stiffness can be found in papers by the Institution of Structural Engineers (1989). The following general observations were made.

Tunnel lining design guide. Thomas Telford, London, 2004

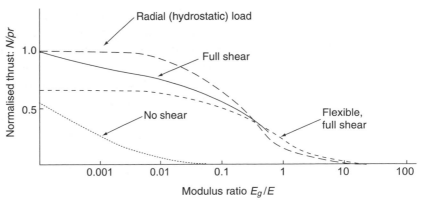

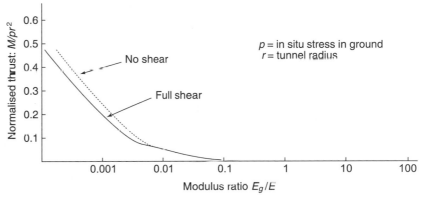

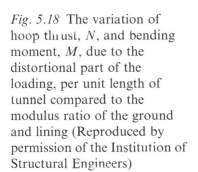

Fig. 5.18 The variation of hoop thrust, N, and bending moment, M, due to the distortional part of the loading, per unit length of tunnel compared to the modulus ratio of the ground and lining (Reproduced by permission of the Institution of Structural Engineers)

N.B. a (truly) flexible lining with no shear at the interface carries no hoop thrust due to distortional loading

N.B. a (truly) flexible lining carries no moments

- *For ground-support modulus ratios of 0.01 or less (see Fig. 5.18), the support will carry as a hoop load, nearly all of the applied load from the overburden.*
- *For ratios greater than 0.01, the reduction in hoop load because of ground–support interaction is substantial.*
- *Distortions of the opening will be determined by the properties of the ground, except at very small modulus ratios.*
- *If, for any reason, the support system is required to limit or resist distortions, the support capacity will have to be considerably increased.*

The modulus ratio is defined as E_g/E, where E_g is the Young's Modulus of the ground and E is the Young's Modulus of the structure. Figure 5.18 shows the variation in hoop thrust and bending moment with modulus ratio, depending on whether or not there is a full shear or zero shear (i.e. full slip) condition at the boundary between the lining and the ground. This is for the distortional part of the ground load. The chart is derived from an analytical solution by Curtis (1976) and Muir Wood (1975). For reference, the hoop load due to the uniform component of the ground load is also shown.

In other words, in tunnels where the modulus ratio is 0.01 or less, the lining is stiff by comparison with the ground. At ratios above 0.01 there is load sharing between the ground and lining, and at ratios above 0.1 the lining can be considered flexible. Others have also examined the effect of varying lining and ground stiffness using different definitions of modulus ratio, for example Peck *et al.* (1972) and the American Society of Civil Engineers (O'Rourke 1984), and found similar results. The flexibility of a lining is increased significantly if the lining consists of segments and there is no moment capacity at the joints between the segments. The effects are shown in Fig. 5.18 by the lines labelled 'flexible'.

Utilising lining flexibility to obtain efficient and economic support systems has been promoted generally by NATM practitioners.

It is fundamentally correct that a confined flexible ring works better than a stiff ring on the basis that there is no advantage in increasing lining thickness where the flexural capacity of the lining is not exceeded. If support is not provided to the ground in time (for example, if a segmental ring is not fully grouted or if a sprayed concrete lining is not closed in a suitable time-frame), then excessive deformation may occur in the ground leading to instability.

5.8.4 Lining distortion

Typical distortions of flexible linings due to ground loading in circular soft ground tunnels are contained in Fig. 5.19. By their jointed nature segmental linings generally conform to the ideal of a confined flexible ring of permanent support. These values can be used in design to check bending moments due to distortion and to assess performance during construction. The distortion is defined as the change in radius, δR, divided by the tunnel radius, R.

Segmental linings generally conform to the ideal of a confined flexible ring since the linings are usually placed as permanent support. The recommended distortion for all situations of not more than 2% of the difference between minimum and maximum diameters is provided in the BTS *Specification for Tunnelling* (British Tunnelling Society, 2000).

Fig. 5.19 Recommended distortion ratios for circular soft-ground tunnels

Soil type	$\delta R/R$
Stiff to hard clays, overconsolidation ratio > 2.5–3.0	0.15–0.40%
Soft clays or silts, overconsolidation ratio < 2.5–3.0	0.25–0.75%
Dense or cohesive sands, most residual soils	0.05–0.25%
Loose sands	0.10–0.30%

5.8.5 Critical strains in the ground

In rock tunnels, where either rock-bolts or thin shotcrete linings are required to prevent surface or key block loosening, distortions are directly related to the stiffness of the rock mass and, for permanent linings, will be considerably less than the values quoted in Fig. 5.19. The heterogeneous nature of many rock masses means that, even with sophisticated numerical modelling techniques, it is often very difficult to predict the actual behaviour accurately.

A useful summary of the difficulties experienced in defining practical limits for assessing the performance, and hence the stability of openings, is presented by Sakurai (1997). He uses the concept of critical strain to define allowable displacements. This is defined for rock masses as

$$\varepsilon_{cr} = (m/n)E_o$$

where E_o is the critical strain for intact materials and m and n are reduction factors of uniaxial strength and Young's Modulus respectively for the rock mass.

In general, the ratio of m/n varies from 1.0–3.0 depending on ground conditions. On most projects an observational approach is used to adjust design input parameters and is combined with

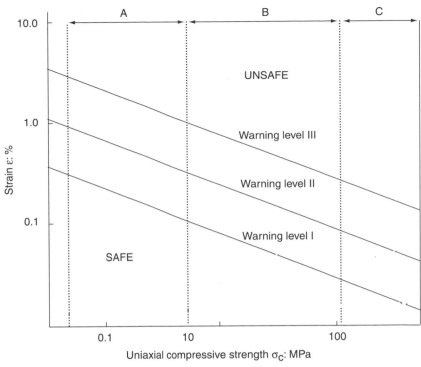

Allowable radial displacements in cm
(for a tunnel of radius — 5.0 m)

Fig. 5.20 Hazard warning levels for assessing the stability of tunnels where strain equals settlement of tunnel crown/ tunnel radius (Sakurai, 1997)

	A	B	C
I	0.3–0.5	0.5–1	1–3
II	1–5	1.5–4	4–9
III	3–4	4–11	11–27

back-analysis from the performance of the structures to allow this ratio to be defined more accurately.

Hoek and Marinos (2000) have looked at the critical strains using the Sakurai approach and concluded that, where the strain exceeds 1% to 2.5%, problems could be experienced with tunnel stability. This is for tunnels in squeezing ground that approximates to an isotropic medium. Design analyses, for example using the Hoek– Brown failure criterion, can be used to assess the likelihood of problems in advance.

Figure 5.20 provides guidance on the likely behaviour of weak rocks, or rocks under high stress, for given rock mass strength/stress

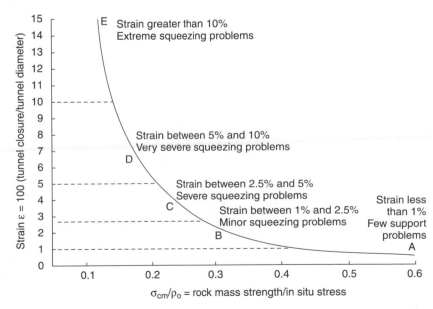

Fig. 5.21 Graph of relationship between strain and degree of unsupported tunnelling difficulty in squeezing rock (after Hoek and Marinos, 2000)

	Strain (%)	Geotechnical issues	Support types
A	Less than 1	Few stability problems and very simple tunnel support design methods can be used. Tunnel support recommendations based upon rock mass classifications provide an adequate basis for design.	Very simple tunnelling conditions, with rock-bolts and shotcrete typically used for support.
B	1–2.5	Convergence confinement methods are used to predict the formation of a plastic zone in the rock mass surrounding a tunnel and of the interaction between the progressive development of this zone and different types of support.	Minor squeezing problems, which are generally dealt with by rock-bolts and shotcrete; sometimes light steel set or lattice girders are added for additional security.
C	2.5–5	Two-dimensional finite element analysis, incorporating support elements and excavation sequence, is normally used for this type of problem. Face stability is generally not a major problem.	Severe squeezing problems requiring rapid installation of support and careful control of construction quality. Heavy steel sets embedded in shotcrete are generally required.
D	5–10	The design of the tunnel is dominated by face stability issues and, while two-dimensional finite analyses are generally carried out, some estimates of the effects of forepoling and face reinforcement are required.	Very severe squeezing and face stability problems. Forepoling and face reinforcement with steel-sets embedded in shotcrete are generally required.
E	More than 10	Severe face instability, as well as squeezing of the tunnel, makes this an extremely difficult three-dimensional problem for which no effective design methods are currently available. Most solutions are based on experience.	Extreme squeezing problems. Forepoling and face reinforcements are usually applied and yielding support may be required in extreme cases.

Fig. 5.22 Relationships between strain and the degree of difficulty in tunnelling for tunnels with no support in squeezing rock (after Hoek and Marinos, 2000)

Tunnel lining design guide. Thomas Telford, London, 2004

ratios. The important conclusion is that below strength/stress ratios of 0.5 the critical strain accelerates rapidly with decreasing strength/stress ratio. Where a rock mass is strongly anisotropic, for example slab failures in strongly bedded sedimentary rocks, it will difficult to determine the critical strain at the point of failure in unsupported ground. As a general recommendation the support should be designed to limit strains to less than 1%.

Figures 5.21 and 5.22 demonstrate the relationship between strain and tunnelling difficulty without support in squeezing rock, reproduced by permission of Tunnels and Tunnelling International copyright Polygon Media Ltd.

5.9 References

AFTES (1990). The design, sizing and construction of precast segments. Association Française des Travaux en Souterrain, Paris (www.aftes.asso.fr).

Anderson, J. M. and Lance, G. A. (1997). The necessity of a risk-based approach to the planning, design and construction of NATM tunnels in urban situations. *Proc. Tunnelling '97 Conf., London.* Institution of Mining and Metallurgy, London, pp. 331–340.

Austin, S. A., Robins, P. J. and Goodier, C. I. (2000). *Construction and Repair with Wet-process Sprayed Concrete and Mortar*, The Concrete Society, Technical Report 56 (Draft), Crowthorne, Berkshire.

Barratt, D. A., O'Reilly, M. P. and Temporal, J. (1994). Long-term measurements of loads on tunnel linings in over-consolidated clay. *Proc. Tunnelling '94 IMMG Conf.* Chapman and Hall, London, pp. 469–481.

Barton, N. (1976). Recent experiences with the Q-system of tunnel support design. *Proc. Symp. on Exploration for Rock Engineering, Johannesburg,* Vol. 1, pp. 107–117, Balkema, Rotterdam.

Bickel, J. O., Kuesel, T. R. and King, E. H. (2002). *Tunnel Engineering Handbook*, 2nd edn. Kluwer Academic Publishers, Dordrecht.

Bieniawski, Z. T. (1984). *Rock Mechanics Design in Mining and Tunnelling.* Balkema, Rotterdam.

Brekke, T. L. and Ripley, B. D. (1987). *Design Guidelines for Pressure Tunnels and Shafts.* Electric Power Research Institute, Final report AP-5273 EPRI.

British Tunnelling Society (2000). *Specification for Tunnelling.* Thomas Telford, London.

Broch, E. (1984). Unlined high pressure tunnels in areas of complex topography. *Water Power and Dam Construction*, November, pp. 21–23.

Clayton, C. R. I., Hope, V. S., Heyman, G., van der Berg, J. P. and Bica, A. V. D. (2000). Instrumentation for monitoring sprayed concrete lined soft ground tunnels. *Proc. Inst. Civ. Eng. Geotechnical Engineering* **143**, July. Thomas Telford, London, pp. 119–130.

Construction Industry Research and Information Association (1999). *The Observational Method in Ground Engineering: Principles and Applications.* CIRIA Report 185, London.

Curtis, D. J. (1976). Discussions on Muir Wood: The circular tunnel in elastic ground. *Géotechnique* **26**, Issue 1. Thomas Telford, London, pp. 231–237.

Deane, A. P. and Bassett, R. H. (1995). The Heathrow Express trial tunnel. *Proc. Inst. Civ. Eng. Geotech. Engng.* **113**, July. Thomas Telford, London, pp. 144–156.

Einstein, H. and Schwartz, W. (1979). Simplified analysis for tunnel supports. *Journal of Geotechnical Engineers.* ASCE, USA.

Golser, J. (1996). Controversial views on NATM, *Felsbau* **14**, No. 2. Austria, pp. 69–75, Verlag Glückhauf GmbH.

Haack, A. (1991). Water leakages in subsurface facilities: required watertightness, contractual matters and method of redevelopment. *Tunnelling and Underground Space Technology* **6**, No. 3. Elsevier, Oxford, pp. 273–282.

Health and Safety Executive (1996). *A Guide to the Work in Compressed Air Regulations 1996 – Guidance on Regulations.* HSE Books, Sudbury, Suffolk.

Health and Safety Executive (1996). *Safety of New Austrian Tunnelling Method (NATM) Tunnels.* HMSO, Norwich.

HM Government (1994). *The Construction (Design and Management) Regulations 1994.* HMSO, London.

Hoek, E. and Brown, E. T. (1980). *Underground Excavations in Rock.* Institution of Mining and Metallurgy, London.

Hoek, E. and Marinos, P. (2000). Predicting tunnel squeezing problems in weak heterogeneous rock masses. *Tunnels & Tunnelling International*, Part 1 November, pp. 45–51, Part 2 December, pp. 33–36.

Huder, J. and Amberg, G. (1970). Quellung in Mergel, Opalinuston und Anhydrit. *Schweitzer Bauzeitung* **88**, pp. 975–980.

Institution of Civil Engineers (1996a). *Sprayed Concrete Linings (NATM) for Tunnels in Soft Ground*. Thomas Telford, London, ICE design and practice guide.

Institution of Civil Engineers (1996b). *Specification for Piles and Embedded Retaining Walls*. Thomas Telford, London.

International Tunnelling Association (1988). Official Report Work Group No. 2: Guidelines for the Design of Tunnels. *Tunnelling and Underground Space Technology* **3**, No. 3. Elsevier Science, Oxford, pp. 237–249.

International Tunnelling Association (2000). Official Report Work Group No. 2: Guidelines for the Design of Shield Tunnel Lining. *Tunnelling and Underground Space Technology* **15**, No. 3. Elsevier Science, Oxford, pp. 303–331.

Isaksson, T. (2002). *Model for Estimation of Time and Cost, Based on Risk Evaluation on Tunnel Projects*. Doctoral thesis, Royal Institute of Technology, Stockholm, Sweden.

IStructE/ICE/IABSE (1989). *Soil–Structure Interaction: The Real Behaviour of Structures*. Institution of Structural Engineers, London.

John, M. (1978). Design of the Arlberg expressway tunnel and the Pfandertunnel in *Shotcrete for Underground Support III*. Austria, pp. 27–43, ASCE.

Lunardi, P. (2000). Design and constructing tunnels – ADECO-RS approach. *T&T International* May special supplement. Miller Freeman, London.

Muir Wood, A. M. (1975). The circular tunnel in elastic ground. *Géotechnique* **25**, Issue 1. Thomas Telford, London, pp. 115–127.

Muller, L. (1978). Removing misconceptions on the New Austrian Tunnelling Method. *Tunnels and Tunnelling* **10**, No. 8, October. Morgan-Grampian, London, pp. 29–32.

Nicholson, D., Tse, C.-M. and Penny, C. (1999). *The Observational Method in Ground Engineering: Principles and Applications*. CIRIA, London, CIRIA Report 185.

Norris, P. (1993). *The Behaviour of Jacked Concrete Pipes during Site Installation*, DPhil thesis. Oxford University, Dept of Engineering Science, Oxford University Press, Oxford.

O'Rourke, T. (ed.) (1984). *ASCE Guidelines for Tunnel Lining Design*. ASCE Geotechnical Division, Reston, Virginia.

Pacher, F. (1977). Underground openings and tunnels in *Design Methods in Rock Mechanics*. ASCE, Reston, Virginia.

Peck, R. B. (1969). Advantages and limitations of the observational method in applied soil mechanics. *Géotechnique* **19**, No. 2. Thomas Telford, London, pp. 171–187.

Peck, R. B., Hendron, A. J. and Mohraz, B. (1972). State of the art in soft ground tunnelling. *Proc. of the Rapid Excavation and Tunneling Conf., New York*, American Institute of Mining, Metallurgical, and Petroleum Engineers (SME), pp. 259–286, Littleton, Colorado.

Pipe Jacking Association (1995). *Guide to Best Practice for the Installation of Pipe Jacks and Microtunnels*. PJA, London.

Powell, D. B., Sigl, O. and Beveridge, J. P. (1997). Heathrow Express – design and performance of platform tunnels at Terminal 4. *Proc. Tunnelling '97 Conf.*, IMMG, London, pp. 565–593.

Proctor, R. V. and White, T. (1946). *Rock Tunnelling with Steel Supports*. Commercial Shearing and Stamping Co., Youngstown, Ohio.

Rabcewicz, L. V. (1969). The New Austrian Tunnelling Method Parts 1, 2 and 3, *Water Power and Dam Construction*, June, pp. 225–229, July, pp. 266–273, August, pp. 297–302.

Rawlings, G., Hellawell, E. and Kilkenny, W. M. (2000). *Grouting for Ground Engineering*. CIRIA, London, CIRIA C514, Funders Report/CP/56.

Ripley, K. J. (1989). *The Performance of Jacked Pipes*, DPhil thesis. Oxford University, Dept of Engineering Science, Oxford University Press, Oxford.

Sakurai, S. (1997). Lessons learned from field measurements in tunnelling. *Tunnelling and Underground Space Technology* **12**, No. 4. Elsevier, Oxford, pp. 453–460. Copyright (1997), with permission from Elsevier.

Seeber, W. (1985a). Power conduits for high head plants – Part I. *International Water Power and Dam Construction*, June.

Seeber, W. (1985b). Power conduits for high head plants – Part II. *International Water Power and Dam Construction*, July.

Soliman, E., Duddeck, H. and Ahrens, H. (1993). Two- and three-dimensional analysis of closely spaced double-tube tunnels. *Tunnelling and Underground Space Technology* **8**, No. 1. Elsevier, Oxford, pp. 13–18.

Szechy, K. (1967). *The Art of Tunnelling*. Akademiai Kiado, Budapest.

Terzaghi, K. (1946). Rock defects and loads on tunnel supports in *Rock Tunnelling with Steel Supports* (eds Proctor, R. V. and White, T.). Commercial Shearing and Stamping Co., Youngstown, USA, pp. 15–19.

Van der Berg, J. P. (1999). *Measurement and Prediction of Ground Movements Around Three NATM Tunnels*. PhD thesis, University of Surrey.

Wittke, W. (1990). *Rock Mechanics: Theory and Application with Case Histories*. Springer-Verlag, Berlin.

6 Theoretical methods of analysis

6.1 Introduction

6.1.1 Purposes

The main purpose of the design analyses is to provide the designer with:

- *an understanding of the mechanisms of behaviour during tunnelling, including the principle risks and where they are located*
- *a basis for producing a robust, safe design*
- *a basis for interpreting the results of monitoring (where applicable).*

The general design procedures for achieving this are summarised in the ITA design process model in Fig. 6.1. Because of the uncertainties surrounding the properties of the ground and the loads on the lining, it is important to understand that tunnel lining design is not a straightforward deterministic process. A design analysis should therefore be regarded as an attempt to provide an indication of the behaviour of the tunnel lining system, rather than the definitive solution.

The ground will usually have the greatest influence on the loads. Very often geotechnical interpretative reports (see Chapter 3) provide a good basis on which to judge ground behaviour. Information from other tunnelling in similar ground is a valuable addition to the results of site investigations.

There is no single method of analysis that can be used for all tunnels. Furthermore, the precision of the available analytical tools is much greater than the reliability and accuracy of the data obtained from site investigations. Therefore designers are obliged to undertake sufficient analyses to understand the sensitivity of the ground-support interaction model to the input parameters. Designers should use a variety of design methods in this process. The methods/tools available are:

- *empirical methods*
- *'closed-form' analytical solutions*
- *numerical models.*

The following sections discuss the available methods of analysis and their strengths and limitations.

6.2 Errors and approximations

All design analyses are only approximations of the real case. In very simple or well-understood cases, for example segmental tunnels in London Clay, the estimations can be quite close to the actual performance. However, this is not always the case and there are six sources of errors in modelling that might lead to poor predictions. After Woods and Clayton (1993), these are:

- *modelling the geometry of the problem*
- *modelling of construction method and its effects*
- *constitutive modelling and parameter selection*
- *theoretical basis of the solution method*
- *interpretation of results*
- *human error.*

Tunnel lining design guide. Thomas Telford, London, 2004

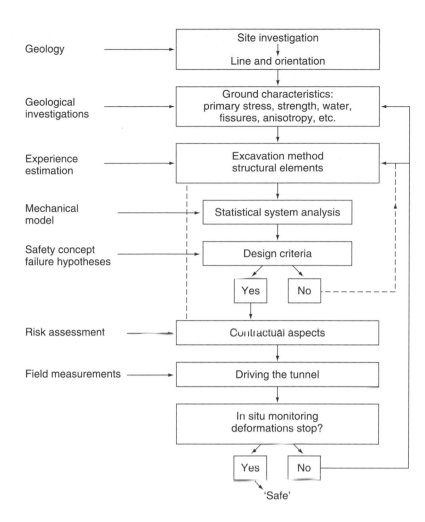

Fig. 6.1 Design approach for rock tunnels (ITA, 1998)

6.2.1 Geometry

Most 'closed-form' analytical solutions are based on a single circular tunnel. However, many transportation projects require complex cross-sectional geometries, junctions and bifurcations. Furthermore the structures may be built in close proximity to each other or to existing structures. An empirical reduction in ground stresses to account for the three-dimensional stress redistribution in the ground ahead of the tunnel is required in simple analytical solutions.

6.2.2 Construction method

A 'wished-in-place' analysis is one that instantaneously creates the tunnel lining in the ground where the stresses are equal to the in situ stresses. This ignores all construction effects. The construction method employed, however, can have a considerable effect on the stress redistribution within the ground and therefore the loads on the lining. Recognition of this fact lies at the heart of tunnel engineering and, while it is explicitly detailed in the NATM design philosophy, this applies to all types of tunnels.

In addition to the construction sequence, the method of excavation (e.g. by earth-pressure-balance TBM, tunnelling in compressed air, or drill and blast) is an important factor. Unfortunately, construction effects are often difficult to quantify, let alone model.

6.2.3 Constitutive modelling

Constitutive modelling is the theoretical simulation of material behaviour. Most materials used for tunnel support are well understood and can be modelled reasonably accurately under the conditions found in tunnels. Exceptions include sprayed concrete at early ages or freshly cast concrete. Ground behaviour is much more complex. At present non-linear ground and support materials are often modelled as linear materials. The ranges of strain, stress, confinement, space and time involved in tunnelling must be considered when choosing a constitutive model. In addition, changes in porewater pressure, drainage and consolidation should be considered.

The significant spatial variability of ground parameters also presents a fundamental problem. Higher quality and more comprehensive site investigations help to improve the level of certainty but sensitivity studies are necessary to assess the influence of variations in ground parameters.

Ultimately, most constitutive models are idealisations of observed behaviour. Parameters are usually based on the moderately conservative or worst credible ground conditions. For more sophisticated constitutive models, determining their parameters and incorporating them into design tools can be extremely difficult.

6.2.4 Theoretical basis

The first step in formulating a 'closed-form' analytical or numerical model of a tunnel system is to decide whether the ground behaves as a single body, that is a continuum, or as a collection of discrete bodies, that is a discontinuum. Sometimes the ground displays elements of both types of behaviour. For example, stiff fissured clay may in general behave as a continuum but there may also be problems of block stability.

When using sophisticated constitutive models a thorough understanding is required of the theory and the formulation of the model in the design tool (e.g. the numerical modelling programme), in order to identify any errors in the results.

6.2.5 Interpretation

Performing calculations is generally fairly simple but converting the results into an effective design for a tunnel support system is more difficult. A thorough appreciation of the limitations of the design methods being used and the project constraints should reveal any inconsistencies. Precedent practice can provide an additional check. Engineering judgement is always required and only experienced tunnel engineers can provide this.

6.2.6 Human error

Errors during calculations can be minimised by an effective checking procedure. Human influences during construction are rarely considered, even though these can be important factors. One exception is the build quality of segmental linings; here the human influence can be considered.

Despite the simplifications in current design methods hundreds of thousands of kilometres of tunnels have been constructed safely throughout the world. Provided that the sources of error in a design are identified and quantified during the design process and managed during construction, lining designs and support systems will be safe and economical.

6.3 Design methods

A summary of the design methods available is contained in Figs 6.2 and 6.3 (for continua and discontinua respectively).

Tunnel lining design guide. Thomas Telford, London, 2004

Method	Source/ example	Material models	2D or 3D	Time effects	Ground water effects[a]	Tunnel shape	Mined/ TBM
Empirical methods							
ADECO-RS	Lunardi, 1997	Based on CCM and numerical analyses	2D	None	–	–	Mined
'Closed-form' analytical methods							
	Muir Wood, 1975 Curtis, 1976 Einstein and Schwartz, 1979 Duddeck and Erdman, 1985	Elastic, plastic, creep	2D	Creep in ground	Some	Circular	Both
CCM	Panet and Guenot, 1982	Elastic, plastic, creep	2D axisym	Creep, timing of support	No	Circular	Both
Bedded beam spring	ITA, 1998	Elastic	2D	None	No	Any	Both
Stability analyses	Mair and Taylor, 1993	Plastic	2D/3D	None	No	Circular	Both
Numerical methods							
FE	e.g. ABAQUS	All	2D/3D	All	Yes	Any	Both
FD	e.g. FLAC	All	2D/3D	All	Yes	Any	Both
FE/BE or FD/BE	e.g. PHASES Hoek et al., 1998	All but BE elastic only	2D/3D	All	Yes	Any	Both

Key				
2D	two-dimensional analysis		Elastic	elastic material behaviour
3D	three-dimensional analysis		Mined	drill and blast or driven heading
axisym	axisymmetric analysis		Plastic	plastic material behaviour
CCM	Convergence–Confinement Method		Some	some of the examples in the category/to some extent
TBM	tunnel boring machine		Creep	creep material behaviour

[a] This column states whether the method provides any information on the effects of or on groundwater, for example porewater changes or consolidation settlements.

Fig. 6.2 Design methods for continua (i.e. 'soft ground' or massive rock)

Method	Source/example	Material models	2D or 3D	Joint orientation	Time effects	Tunnel shape	Mined/ TBM
Empirical methods							
RMR	Bieniawski, 1984	–	–	Yes	None	Any	Mined
Q-system	Barton et al., 1974	–	–	No	None	Any	Mined
RMi	Palmstrom, 1996	–	–	Yes	None	Any	Mined
Analytical methods							
Stability analyses	Barrett and McCreath, 1995	Plastic	2D	Some	None	N/A	Mined
Numerical methods							
DE	UDEC/3DEC	All	2D/3D	Yes	Creep	Any	Both
BE	–	Elastic	2D/3D	Yes	None	Any	Both

Key				
2D	two-dimensional analysis	Mined	drill and blast or driven heading	
3D	three-dimensional analysis	Plastic	plastic material behaviour	
TBM	tunnel boring machine	Creep	creep material behaviour	
Elastic	elastic material behaviour	N/A	not applicable	

Fig. 6.3 Design methods for discontinua (i.e. jointed rock masses)

6.3.1 Empirical methods

Empirical methods are based on assessments of precedent practice and generally have a successful track record in rock tunnels. Ideally the support recommendations have been 'calibrated' against actual performance for a wide range of tunnelling conditions and tunnel sizes. Some soft ground empirical rules have also been established but these tend to be based on local experience rather than being universally applicable.

The most frequently used empirical design methods are the RMR – Rock Mass Rating – (Bieniawski, 1994) and the Q systems (Barton *et al.*, 1974), see Figs 6.4 and 6.5. These employ a combination of parameters such as the strength of the rock, its quality by using RQD values (Barton, 1999), joints, and number of sets, frequency, spacing and condition, and groundwater conditions to produce a rock mass classification. Using the product of these parameters, the support measures required can be determined from design charts or tables. A measure of their success has been the number of publications amplifying specific aspects and developments of the original format. These methods, and others, are reviewed in Hoek and Brown (1980) and Hoek *et al.* (1998).

The Q system has even been extended to provide estimates of TBM advance rates in rock via Q_{TBM} (Barton, 1999).

Empirical methods have been used by Hoek *et al.* to formulate and quantify an empirical failure criterion for rock masses. Subsequently, this has been upgraded and modified and the

---

Design approach for rock tunnels

1. Classify rock (using RMR or Q-system classifications – ideally both)

2. Use RMR or Q-system design charts to make preliminary estimate of support

3. Consider overall stability and possible stress instability
    (i) Consider in situ stress
    (ii) Consider excavation geometry
    (iii) Consider wedge stability

4. Consider excavation sequence and timing of support

5. Check rock-bolt design against other empirical rules

6. Draw up support and tunnel to scale in order to visualise the support and possible failure mechanisms (Note: this is difficult to do when it is a three-dimensional problem)

Fig. 6.4 General procedures for applying rock mass classification systems

---

RMR	Q-system
Applicable to tunnels and mines	Applicable to tunnels only
Considers orientation of joint sets	No explicit consideration of joint orientation
Provides information on stand-up time	Does not provide stand-up time
Applicable for permanent support only	For temporary and permanent support
Provides some information on lining loads	Provides some information on lining loads
Does not consider stress/strength ratio	Stress/strength ratio considered
Support chart is not up-to-date	Support chart is up-to-date
Does not consider time-dependent ground behaviour or squeezing rock behaviour	Does not consider time-dependent ground behaviour or squeezing rock behaviour

Key: $RMR = 9 \ln(Q) + 44$ approximately (Bieniawski, 1994)

Fig. 6.5 Limitations of RMR and Q-systems

Tunnel lining design guide. Thomas Telford, London, 2004

GSI – Geological Strength Index – system introduced to compensate for inconsistencies in weak rocks and shallow tunnels that became apparent with the earlier approaches.

The RMR and Q systems have been adapted by Serafim and Perriera (1983) to derive a correlation between rock mass quality and rock mass modulus. While the classification is useful, care has to be taken at the low end of the RMR and Q ratings where the predicted modulus values are sensitive to small variations in the ratings. Unfortunately, these are also the ranges that are most important when evaluating the performance of key excavations such as portals and optimising support requirements in poor ground.

Classification systems such as those of Terzaghi (1946), and Wickham, Tiedemann and Skinner in 1972 (see Hoek and Brown, 1980 for both) focus on US tunnelling practice and reflect the widespread use of steel sets, lagging and rock-bolts. While steel sets can be very effective, systems that rely on some unrestrained deformations to fully activate the load bearing capacity often run the risk of allowing loosening around the excavation perimeter. Better control of deformations is achieved by the application of sprayed concrete to embed steel ribs and directly support the ground.

The strength of empirical methods generally lies in their simplicity, as well as speed and economy of use. Since the approaches are based on practical experience, they are particularly suitable for feasibility studies at the concept design stage. Also, because of the difficulty of investigating and modelling ground behaviour in complex rock masses, the averaging of rock mass properties and specifying rock support classes at the detailed design stage provides a useful and necessary basis for managing support selection during construction. During construction, the support can be chosen from the pre-designed support classes on the basis of the rock mass classification of the exposed face.

Experience over many years has shown that the methods are generally successful when implemented by experienced tunnel engineers or engineering geologists. However, potentially there are disadvantages that should be taken into consideration. They (after Riedmuller and Schubert, 1999) are as follows:

- **Extrapolating to situations that differ from the original data set**. This could lead to support recommendations that are incompatible with the predicted ground behaviour. This is of more concern in heavily fractured weak to very weak rock.
- **The factor of safety in the design is unknown.**
- **There is little or no guidance on the timing of support installation**.
- **There is no consideration of the effects of adjacent structures, either man-made** (e.g. tunnels) **or natural** (e.g. faults).

It is important to understand these limitations and assess whether more sophisticated methods of analysis are required.

Successful application of empirical methods requires regular inspection and monitoring of the tunnel during construction. This is an observational process since decisions are based on a continuous assessment of tunnelling conditions. The 'Observational Method' was presented by Peck (1969) and is best defined in Everton (1998) as:

A continuous, managed and integrated process of design, construction control, monitoring and review, which enables previously designed modifications to be incorporated during or after construction as appropriate. All these aspects have to be demonstrably robust. The objective is to achieve greater overall economy without compromising safety.

This is not 'design as you go'. A robust design is drawn up in advance and it is recognised that this may be altered as construction progresses.

6.3.2 'Closed-form' analytical methods

Soil deformations ahead of the face, stress relief prior to the installation of support and soil-structure interaction determine the stresses and strains in tunnel linings. While some analytical solutions can model this, they are unable to model the full complexity of a tunnel during construction. Specifically, they are generally two-dimensional idealisations that assume the ground is a homogeneous continuum and the tunnel is circular (see Fig. 6.2).

Analytical design tools can be used for designing adjacent tunnels using the principle of superposition but this assumption may significantly underestimate the interaction of the tunnels, especially if the distance between the two tunnels is less than two clear diameters (Szechy, 1967). Also, these models do not make any allowance for construction loads (e.g. from rams in a TBM) or the timing of support placement, with the exception of the Convergence–Confinement Method.

However, they have a proven track record in soft ground such as London Clay and are still widely used for dimensioning tunnel linings in simple cases. A comprehensive review of analytical models is presented by Duddeck and Erdman (1985).

In terms of the range of analytical solutions available, various methods exist for determining the stresses around a hole in an elastic or elasto-plastic homogeneous half-space. Various pressure distributions have been proposed to derive the stresses in the lining, for example Terzaghi or Protodiakonov (see Szechy, 1967). These are of limited use in determining the loads on linings because they consider the ground alone. More useful analytical methods are presented below.

6.3.2.1 Continuum analytical models Commonly used continuum analytical models, also referred to as 'closed-form' solutions, include those proposed by Muir Wood (1975), Einstein and Schwartz (1979) and Duddeck and Erdman (1985). All of these models are based on excavation and lining of a hole in a stressed continuum. In general these models yield similar results for normal forces for the same input parameters but the predicted bending moments may differ significantly.

Most 'closed-form' analytical solutions assume plane stress, an isotropic, homogeneous elastic medium and an elastic lining for a circular tunnel, although the Curtis–Muir Wood solution has been extended by Curtis (1976) to viscoelastic ground. The assumption that the lining is installed immediately after the tunnel is excavated tends to overestimate the loads and hence judgement is required in deciding the proportion of the original in situ stresses to apply to the linings.

Options include applying a reduction factor to the full applied ground stress; any stress relief depends on the ground conditions and the method of construction. This reduced stress can be assumed at 50–70% if the depth to tunnel axis is greater than three diameters (Duddeck and Erdman, 1985). Alternatively, the K_0 value can be set at less than 1.0 to simulate actual behaviour, that is the tunnel 'squats', to match the observed behaviour of segmental tunnels in soft ground.

These models also assume that the ground is a semi-infinite medium and therefore they should only be used for tunnels where the axis is deeper than two tunnel diameters below the surface. Duddeck and Erdman recommend that full bonding at the ground–lining interface be assumed for the continuum models listed above. Most analytical solutions are formulated in total stresses, although there are some that can be used to estimate changes in pore pressures (Mair *et al.*, 1993).

The benefit to the designer is that the methods are simple and quick to use. Information is provided on the normal forces, bending moments and deformations and several methods should be applied with a range of input parameters to determine the sensitivity of the lining designs to variations in ground conditions.

6.3.2.2 Convergence–Confinement Method (CCM)

The Convergence–Confinement Method (Panet and Guenot, 1982 and Hoek and Brown, 1980) is able to predict the deformation of the ground for a wide range of ground conditions and tunnel support measures. As with most design approaches, it has developed over time and now includes the effects of plasticity according to the Mohr–Coulomb or Hoek–Brown yield criteria; creep in the ground; gravity effects; the timing of support via the geometric delay parameter; support type (sprayed concrete, concrete, steel sets and rock-bolts) and mined or TBM-driven tunnels (Eisenstein and Branco, 1991).

For more complex applications of the method, a numerical solution is required. While the assumption of axisymmetry, for ground behaviour and tunnel geometry, represents a significant advantage since it permits the stress relief ahead of the face to be modelled, it is also a significant limitation. Hence, the method is valid only for circular cross-sections at medium depth where K_0 is close to 1.0 and the tunnel is constructed using full-face excavation techniques. In addition no information is given on the distribution of bending moments in the lining.

6.3.2.3 Limit-equilibrium methods

Rock support systems can be designed using limit-equilibrium methods of analysis. The support requirements for individual wedges can be calculated by hand or using programs such as UNWEDGE to provide a graphical presentation of the wedge geometry and assess the distribution of support (Hoek and Brown, 1980).

Similar calculations for a variety of failure mechanisms can be performed to determine the required thickness of sprayed concrete between rock-bolts. The governing failure mechanism is generally loss of adhesion to the rock, followed by flexural failure of the sprayed concrete, Barrett and McCreath (1995).

6.3.2.4 Bedded-beam-spring models

These simulate a tunnel lining as a beam attached to the ground, which is represented by radial and tangential springs, or linear elastic interaction factors, to allow for ground support interaction. The stiffness of the springs can be varied to model conditions at the tunnel extrados from 'no slip' to 'full slip', and different load combinations can be modelled. Relationships exist for determining the spring stiffness from standard ground investigation tests.

Despite the fact that these models tend to underestimate the beneficial effects of soil–structure interaction, and cannot consider shear stresses in the ground itself, the results can sometimes agree well with those from continuum analytical models (O'Rourke, 1984). Also see Duddeck and Erdman (1985) for further details of the closed form solution of Schulze and Duddeck's partially bedded-beam model and a comparison with continuum analytical models.

One of the drawbacks with this method of analysis is the lack of information on movements in the ground and therefore two-dimensional numerical models have tended to replace bedded-beam models. It is also difficult to determine the spring stiffnesses.

6.3.3 Numerical modelling

In contrast to the design methods already outlined, numerical analyses, such as those using the finite element (FE) and finite difference (FD) methods, offer the ability to model explicitly complex structures, including adjacent structures, different geological strata, complex constitutive behaviour, transient and dynamic loading, and construction sequences. This provides an unparalleled capability for simulating ground-support interaction and has led to numerical methods replacing other methods of analysis. Commercially available programs for numerical analysis offer a wide range of constitutive models and elements, and it is possible to model almost any situation. However, as noted earlier, lining design is not a deterministic process and the results should be assessed in the context of the quality of the site investigation and the estimated range of geomechanical properties.

Most computer programs receive input at one end and produce 'the answer' at the other. Within the accuracy of the solution algorithm, a computer program will produce an answer that is correct according to the input data. In reality the 'answers' are still only estimates of how the ground and tunnel lining interact. Numerical modelling remains subject to the six sources of error listed in Section 6.2. For this reason, they are often considered more useful as a tool to investigate mechanisms of behaviour rather than as a means of obtaining precise predictions about tunnel performance (Coetzee *et al.*, 1998).

The methods of analysis that are available, and the strengths and weaknesses and practical problems associated with numerical modelling, are discussed below. Despite significant improvements on the hardware side in recent years, simplifications in numerical analyses are still driven by the limitations in computing power.

6.3.3.1 Methods of numerical analysis
A variety of two- and three-dimensional modelling programs are available. The choice of program depends on whether the ground can be modelled as a continuum or whether the influence of discontinuities, for example faults, bedding surfaces, joints, shear zones, etc., requires an assessment of independent block movements.

- **Soft ground** Soft ground is normally considered as a continuum and hence finite element (FE), hybrid finite element/boundary element (FE/BE) or finite element/finite difference (FE/FD) and finite difference (FD) methods can be applied.
- **Rock** Jointed rock masses are discontinua and often can be modelled realistically using discrete element (DE) and boundary element (BE) methods. Discrete element methods include distinct

Tunnel lining design guide. Thomas Telford, London, 2004

element programs in which the contacts between elements may deform and discontinuous deformation analysis programs in which the contacts are rigid. In addition by means of interface elements, a small number of discontinuities can be modelled in FE and FD models, but DE is required when modelling intersecting joints and larger numbers of discontinuities.

In more complex ground conditions, the tunnel may pass through discontinuous and continuous media in close succession, for example a tunnel in jointed rock intersecting a fault filled with clay gouge and highly sheared rock. A few programs can combine methods, for example Itasca's FD and DE programmes can be coupled to form a 'composite' model. Often this is a means of reducing the mesh size and analysis time.

6.3.3.2 Finite element and finite difference The most commonly used methods are FE and FD. The process of building a model with these methods is essentially the same and the end products are often very similar (Coetzee *et al.*, 1998). The object to be analysed is represented by a mesh of many elements or zones, in a process of discretisation. The material properties, material behaviour, boundary conditions and loads are assigned to the model and the problem solved.

In FE a stiffness matrix is assembled for the whole mesh in order to relate the displacements to the stresses. These vary in a prescribed manner within each element. The matrix is then solved using standard matrix reduction techniques, in a so-called 'implicit' solution technique.

In the FD method, the 'dynamic relaxation' solution technique is used. Newton's Law of Motion is expressed as a difference equation and is used to relate explicitly the unbalanced forces at each integration point in the mesh to the acceleration of the mass associated with that point. For a very small time-step the incremental displacements can be calculated. In static mechanical problems this time step is fictitious, i.e. it is not related to real time. The incremental displacements are used to calculate a new set of unbalanced forces (from the constitutive relationships). This calculation step is repeated many times for each integration point in the mesh, in a 'time-marching' method, until the out-of-balance force has reduced to a negligible value, i.e. equilibrium has been reached for a statical problem. More integration points are required in an FD rather than an FE model because FD uses constant strain zones.

6.3.3.3 Discrete element and boundary element In the DE method the individual blocks in a rock mass are modelled and the elements may move and rotate, depending on the movement of adjacent elements. Either FE or FD is used to model the constitutive behaviour within the elements.

In the BE method the surface of an object is divided into elements, which are modelled mathematically as infinite continua. A more detailed description of all of these numerical methods can be found in Hoek *et al.*, 1995.

6.3.3.4 General application As a general comment, the advantage of the DE method is its ability to model, in an idealised form, the individual blocks making up the rock mass. However, setting up the model and predicting displacements is time-consuming and difficult to undertake, especially for three-dimensional models.

The BE method is rarely used on its own because it uses only elastic elements and excavation sequences cannot be modelled (Schwciger and Beer, 1996).

Usually the FE method, using implicit solution techniques, can solve linear and moderately non-linear problems much faster than the FD method, using explicit solution techniques. However, for more complex non-linear problems the number of iterations and subdivisions of the load into load increments required by the FE method means that the FD method often performs better. The explicit, time-marching solution method removes the need for subdivision of the load and permits both material and geometric non-linearity, for example creep or large strain behaviour, to be modelled easily (Hoek *et al.*, 1998). By comparison, re-meshing is required for FE models in large strain behaviour. Physical instability is also easier to model and detect using the FD method.

6.3.4 Modelling geometry

For analytical purposes a tunnel is normally represented as a two-dimensional model assuming transverse plane strain and axisymmetry. Alternatively one could use a longitudinal plane strain model but this implies the excavation of an infinitely wide slot in the ground. This is justified on the grounds that the transverse stress and strain distributions for a tunnel of reasonable length in homogeneous ground corresponds to a two-dimensional plane strain condition at distances greater than two or three diameters from the point at which the tunnel ring is closed. The actual stress redistribution, particularly at the advancing face, is three-dimensional so, strictly speaking, the best practice would be modelling in three dimensions when considering the lining near the face. Such three-dimensional models are complex to construct, time-consuming to analyse and difficult to interpret.

To allow for the three-dimensional stress redistribution effect in 2D analyses, commonly used techniques include: percentage unloading; volume loss; progressive softening; and gap parameter methods (Van der Berg, 1999). In practice, all of these approaches permit a certain amount of deformation in the ground and the designer is required to estimate the percentage of the total deformation that best represents the actual performance. This can be on the basis of precedent practice, using empirical methods (e.g. Macklin, 1999), using analytical solutions, for example the Convergence–Confinement Method, or by doing a trial run with a 3D mesh. Clearly this is a critical input parameter and, with undrained elastic analyses, there is a risk of 'prejudging' performance in that the volume of the predicted surface settlement curve correlates with the volume loss at the tunnel.

Whichever procedure is used, the associated stress redistribution leads to a reduction in applied pressure before the lining is introduced and stress levels and distributions should match more closely the expected lining performance. If computing power is limited, there is a risk with three-dimensional models that the simplifications required to reduce the size of the model are more significant than the potential error in the correction for three-dimensional effects in a two-dimensional model.

6.3.5 Discretisation

The fineness of the mesh affects the accuracy of the results. There should be more elements (or zones) where there are large stress

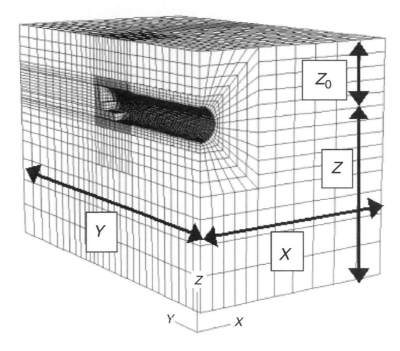

Fig. 6.6 Typical mesh for a numerical model

X	Y	Z	Source
$15 \times R = 3.5 \times Z_0$	$15 \times R$	$3 \times Z_0$	Van der Berg, 1999
$3 \times Z_0$	–	$2 \times Z_0$	Gunn, 1993

where R = tunnel radius and Z_0 – depth from surface to tunnel axis

Fig. 6.7 Mesh dimensions

gradients or where a higher resolution is required (e.g. the centre of the mesh). Aspect ratios for elements and zones should be less than 5:1 and where possible symmetry should be exploited to reduce the size of the mesh.

6.3.5.1 Boundary conditions For lining design, the main interest is the tunnel lining and the ground in the immediate vicinity of the tunnel. The boundaries of the model should be far enough away not to influence the results of the analysis, see Figs 6.6 and 6.7.

Certain methods, for example 'Phases', avoid this problem by using boundary elements or infinite finite elements on the sides of the FE mesh (Schweiger and Beer, 1996).

Fixities of stress and displacement are applied at the boundaries to prevent the model moving as a rigid body and to maintain the appropriate boundary conditions during the analysis. If displacement is fixed at the boundaries, both stresses and displacements are likely to be underestimated. If the stresses at the boundary are fixed, both stresses and displacements are likely to be overestimated. The 'true' solution lies somewhere in between. A small difference between the results for these two sets of boundary conditions is an indication of a well-constructed model.

6.3.5.2 Element types Numerical problems, such as 'volumetric locking', can be encountered with low-order finite elements in elastoplastic media and this has prompted the development of more sophisticated elements (Groen and de Borst, 1997).

6.3.6 Modelling construction processes
The method of construction, and associated procedures, are fundamental in determining the loads on the lining and should be modelled as closely as possible in any design analysis. This particularly applies

to sprayed concrete lined tunnels with complex excavation sequences. Other processes and their effects (e.g. compensation grouting, the pressures induced around the head of an EPB machine, or air losses from a tunnel under compressed air) are as difficult to model. Certain structural elements or processes, for example forepoling or face dowels, can only be modelled explicitly in 3D analyses.

6.3.7 Constitutive modelling

Since tunnelling is a ground–support interaction problem, the constitutive models for the ground and lining are important but this importance depends on what is of most interest in an analysis. For example, it has been suggested that the predictions of surface settlement do not depend greatly on the constitutive model of the lining (Bolton et al., 1996). The following comments on models are worth noting for guidance.

6.3.7.1 Geotechnical model Most numerical analyses are performed using simple ground models, either linear elastic or elastic–perfectly plastic despite the fact that they are not realistic representations of the ground's behaviour. For that reason they do not yield good predictions of ground movements. It has also been found necessary in certain cases to include factors such as non-linear stress–strain behaviour, (Gunn, 1993), and anisotropy (Lee and Rowe, 1989).

The initial state of stress is also a major influence on the stresses in the ground and lining (Addenbrooke and Potts, 1996). Apart from increasing the runtime of the analyses, the use of more sophisticated constitutive models can be problematic since their parameters may be difficult to determine from the site investigation data. This can be overcome by careful planning of the site investigation (Potts and Zdravkovic, 1999).

In the crown and invert the soil near the tunnel undergoes triaxial extension while, at the axis, the stress path lies between triaxial compression and extension. The stiffness of the ground may vary significantly depending on the load path (Lee and Rowe, 1989).

The hydraulic conductivity of the ground and lining, hydrostatic pressures and the initial state of stress, have an important influence on the loads in a lining. For materials such as clay with a low permeability in the short term, and this usually covers the period required to install the primary lining, one can assume undrained behaviour. More generally effective stress analyses should be modelled since the strength of the ground is heavily dependent on the porewater pressures. In the long term, if the excess pore pressures due to construction are allowed to dissipate, that is the lining is not watertight, changes in effective stress will lead to consolidation and additional loads on the lining. In ground that has a high permeability, either the hydrostatic pressure will remain unchanged, if the tunnel lining is watertight, or seepage forces will act on the lining, if the tunnel acts as a drain. The actual behaviour is often complex and as the model complexity increases so does the difficulty in interpreting its results. Performing coupled stress–fluid flow analyses, i.e. consolidation, may improve the prediction of settlements but the pattern of excess pore pressures depends heavily on the geotechnical model and how the tunnel lining is modelled (Addenbrooke and Potts, 1996).

6.3.7.2 Lining model Compared to the ground, the tunnel lining usually consists of materials that are much more uniform, relatively simple in their behaviour and better understood, such as mature concrete or sprayed concrete and steel. However, in the case of sprayed concrete used for immediate support, its early age properties such as strength and stiffness change considerably during the construction period. Creep and shrinkage may be significant influences on the stresses and strains in the lining. Often a reduced value of the stiffness of the sprayed concrete, that is a Hypothetical Modulus of Elasticity (HME) is used to account for these effects as well as the stress distribution ahead of the face in 2D analyses. The HME is essentially an empirical correction factor that adjusts the lining modulus to match the expected deformations. However, there is no rigorous basis for determining the HME, although Pottler (1985) proposes a largely empirical method.

Where numerical analyses are fundamental to the design, the lining model should account for all components of the lining, including the actual behaviour of its constituents in both the short term and long term. Possible variability of construction and quality control defects, for example ovalisation in segmental linings and variations in thickness and geometry of sprayed concrete linings, should also be considered, although it is not always feasible to model them explicitly.

For example, with reinforcement it would be preferable, in theory, to model the individual bars in a lining but this would require enormous effort. One can account for such components by modifying the bulk properties of the lining.

6.3.8 Validation

Each numerical modelling program is encoded in its own way and, because of different solution algorithms or element types, this can lead to variations in results. The more complex the numerical modelling program, the harder it is to validate. It is advisable to start any set of numerical analyses with some trial runs using a simplified mesh and linear elastic models. This will help to identify problems with the mesh geometry and modelling of construction processes. Where possible the results should be compared to an analytical solution (see Section 6.3.2) to check that the model is functioning correctly. Following these procedures will assist in reducing human error, one of the main concerns when using numerical methods.

Once the basic model has been validated, more complex models can be analysed and the results compared with existing field data and analytical solutions where possible; always assuming that the field data are reliable and of a high quality. All of the significant input parameters should be varied systemically over their credible ranges in a series of runs, so that the influence of each parameter on the results can be determined.

As stated earlier, in all tunnel designs, because of the uncertainty and complexity involved in the problems, more than one design method should be used to provide independent checks.

6.3.9 Advances in numerical analyses

Recent developments in this field include the use of neural networks, stochastic methods and a more widespread use of back analysis.

6.3.9.1 Neural networks In very simple terms, neural networks are computational routines, which can 'learn' from experience.

Once a neural network has been trained to perform a task using existing field data, it can be used to predict results for new cases. Neural networks have been used to predict surface settlements (Ortigao and Shi, 1998).

6.3.9.2 Stochastic methods (Probability Theory) These may be used to account for the inherent variability of the ground. Key input parameters, such as soil stiffness, often vary randomly but this can be idealised as a probability distribution, for example as a normal (or Gaussian) distribution centred on a mean value. A solution for a problem can be found by repeatedly sampling the distribution of input parameters, and solving the problem for those inputs. The results of all these analyses form a probability distribution of the output, for example the factor of safety against collapse. This provides a more comprehensive picture of the behaviour than one would obtain from a few analyses with the best estimates of input parameters. However, it is difficult and time-consuming to apply to complex problems such as the numerical analyses of tunnels. An overview of this method and sampling techniques, such as the Monte Carlo or Latin Hypercube techniques, together with a worked example, can be found in Hoek *et al.*, 1995.

6.3.9.3 Back-analysis The aim of back-analysis is to calibrate the design model with measurements taken during construction. This is a 'trial-and-error' process that systematically varies the input parameters until a 'best' fit with actual performance is achieved. In the case of a long rock tunnel this process assists in refining the design model in order to improve predictions for remaining tunnel sections. The results will not allow for unforeseen local factors but should provide a sound basis for assessing general performance.

6.3.10 Physical modelling

Physical modelling plays an important role in the design of tunnels, not least because extreme conditions, such as collapse, can be examined safely. However, its direct use in the design of linings has been limited to date.

Centrifuge modelling is most widely used to examine collapse mechanisms and movements in the ground, for example Taylor (1998). As the technology has developed, the centrifuge models have become more realistic and can now replicate actual tunnel construction and related processes, for example compensation grouting. For practical reasons few models have examined the behaviour of the tunnel lining itself.

Large-scale models of tunnel linings have been constructed and tested to examine behaviour under working and collapse loads (e.g. as part of a recent Brite Euram project, BRITE–EURAM, 1998). Trial tunnels provide the most readily accessible and realistic data on the performance of tunnel linings, although at considerable expense. Examples from the UK include the Kielder experimental tunnel (Ward *et al.*, 1983), the trial tunnels for the Heathrow Express (Deane and Bassett, 1995) and the Jubilee Line Extension (Kimmance and Allen, 1996).

On a general note, the results from monitoring during and after construction bolster the understanding of tunnel behaviour and the long-term loads on linings in soft ground, and are essential both to enhance design methods and to calibrate numerical models (e.g. Barratt *et al.*, 1994).

6.4 Recommendations on design methods

Tunnel lining design is a challenging task, not least because of the variability of the ground. Therefore it should be approached as an iterative process, in which the designer may use a variety of design methods, in order to gain an appreciation of how the ground and lining are likely to interact. From that the support required can be determined to maintain safety both in the short- and long-term and to satisfy project requirements. Sound engineering judgement underpins this process.

Empirical, 'closed-form' analytical and numerical design methods exist. Each method has its own strengths and limitations. These should be borne in mind when interpreting the results of design calculations. It is recommended that several design methods be used when designing a lining, since the other design methods will provide an independent check on the main design method.

6.5 References

Addenbrooke, T. I. and Potts, D. M. (1996). Twin tunnel construction – ground movements and lining behaviour in *Geotechnical Aspects of Underground Construction in Soft Ground* (eds Mair, R. J. and Taylor, R. N.). Balkema, Rotterdam.

Barratt, D. A., O'Reilly, M. P. and Temporal, J. (1994). Long-term measurements of loads on tunnel linings in overconsolidated clay. *Proc. Tunnelling '94 Conf.* Chapman and Hall, London, pp. 469–481.

Barrett, S. V. L. and McCreath, D. R. (1995). Shotcrete support design in blocky ground: towards a deterministic approach. *Tunnelling and Underground Space Technology*, Elsevier, Oxford.

Barton, N. (1999). TBM performance estimation in rock using Q_{TBM}. *Tunnels & Tunnelling*, September, pp. 30–34.

Barton, N. R., Lien, R. and Lunde, J. (1974). Engineering classification of rock masses for the design of tunnel support. *Rock Mechanics* **6**, No. 4, pp. 189–239.

Bieniawski, Z. T. (1994). *Rock Mechanics Design in Mining and Tunnelling*. Balkema, Rotterdam.

Bolton, M. D., Dasari, G. R. and Rawlings, C. G. (1996). Numerical modelling of a NATM tunnel construction in London Clay in *Geotechnical Aspects of Underground Construction in Soft Ground* (eds Mair, R. J. and Taylor, R. N.) pp. 491–496 CIRIA.

Brite Euram (1998). *New Materials, Design and Construction Techniques for Underground Structures in Soft Rock and Clay Media.* BRE-CT92-0231 Final Technical Report, Mott MacDonald Ltd.

Coetzee, M. J., Hart, R. D., Varona, P. M. and Cundall, P. A. (1998). *FLAC Basics*. Itasca Consulting Group, Minnesota.

Curtis, D. J. (1976). Discussions on Muir Wood. The circular tunnel in elastic ground. *Géotechnique* **26**, No. 1. London, pp. 231–237.

Deane, A. P. and Bassett, R. H. (1995). The Heathrow Express trial tunnel. *Proc. ICE Geotechnical Engineering* **113**, July, pp. 144–156.

Duddeck, H. and Erdman, I. (1985). On structural design models for tunnels in soft soil. *Tunnels & Deep Space*.

Einstein, H. and Schwartz, W. (1979). Simplified analysis for tunnel supports. *Journal of Geotechnical Engineers.* ASCE, USA.

Eisenstein, Z. and Branco, P. (1991). Convergence–Confinement Method in shallow tunnels. *Tunnelling and Underground Space Technology* **6**, No. 3. Elsevier, Oxford, pp. 343–346.

Everton, S. (1998). Under observation. *Ground Engineering* May, pp. 26–29.

Groen, A. E. and de Borst, R. (1997). Three-dimensional finite element analysis of tunnels and foundations. *Heron* **42**, Issue 4, pp. 183–214.

Gunn, M. J. (1993). The prediction of surface settlement profiles due to tunnelling. *Predictive Soil Mechanics*, Thomas Telford, London.

Hoek, E. and Brown, E. T. (1980). *Underground Excavations in Rock*, IMMG, London.

Hoek, E., Kaiser, P. K. and Bowden, W. F. (1995). *Support of Underground Excavations in Hard Rock*. Balkema, Rotterdam.

International Tunnelling Association (1998). Working group on general approaches to the design of tunnels: Guidelines for the design of tunnels. *Tunnelling and Underground Space Technology* **3**, Elsevier Oxford, pp. 237–249.

Kimmance, J. P. and Allen, R. (1996). NATM and compensation grouting trial at Redcross Way in *Geotechnical Aspects of Underground Construction in Soft Ground* (eds Mair, R. J. and Taylor, R. N.), Balkema, Rotterdam.

Lee, K. M. and Rowe, R. K. (1989). Deformations caused by surface loading and tunnelling: the role of elastic anisotropy. *Géotechnique* **39**, Issue 1, pp. 125–140.

Lunardi, P. (1997). The influence of the rigidity of the advance core on the safety of tunnel excavations. *Gallerie*, No. 52, July, Italy.

Macklin, S. R. (1999). The prediction of volume loss due to tunnelling in over-consolidated clay based on heading geometry and stability number. *Ground Engineering* April, pp. 30–34.

Mair, R. J. and Taylor, R. N. (1993). Prediction of clay behaviour around tunnels using plasticity solutions in *Predictive Soil Mechanics*, Thomas Telford, London, pp. 449–463.

Mair, R. J., Taylor, R. N. and Bracegirdle, A. (1993). *Géotechnique* **43**(2), Thomas Telford, London, pp. 315–320.

Muir Wood, A. M. (1975). The circular tunnel in elastic ground. *Géotechnique* **25**, Issue 1. Thomas Telford, London, pp. 115–127.

O'Rourke, T. D. (1984). *Guidelines for Tunnel Lining Design*. Technical Committee 011 Tunnel Lining Design of the Underground Technology Research Council of the ASCE Technical Council.

Ortigao, J. A. R. and Shi, J. (1998). Settlement monitoring. *Tunnels & Tunnelling International* December, pp. 30–31.

Palmstrom, A. (1996). Characterizing rock masses by the RMi for use in practical rock engineering. *Tunnelling and Underground Space Technology* Part 1: **11**, No. 1 and Part 2: **11** No. 3. Elsevier, Oxford.

Panet, M. and Guenot, A. (1982). Analysis of convergence behind the face of a tunnel. *Proc. Tunnelling '82 Conf*. IMM, pp. 197–204, London.

Peck, R. B. (1969). 9th Rankine Lecture: Advantages and limitations of the Observational Method in applied soil mechanics. *Géotechnique* **19**, No. 2, Thomas Telford, London, pp. 171–187.

Pottler, R. (1985). Evaluating the stresses acting on the shotcrete in rock cavity constructions with hypothetical modulus elasticity. *Felsbau*, **3**, No. 3, pp. 136–139.

Potts, D. M. and Zdravkovic, L. (1999). *Finite Element Analysis in Geotechnical Engineering: Theory*. Thomas Telford, London.

Riedmuller, G. and Schubert, W. (1999). Critical comments on quantitative rock mass classifications. *Felsbau* **17**, No. 3, pp. 164–167.

Schweiger, H. and Beer, G. E. (1996). Numerical simulation in tunnelling. *Felsbau* **14**, pp. 87–92, Verlag Glückauf GmbH.

Serafim, J. L. and Perriera, J. P. (1983). Considerations on the geomechanical classification of Bieniawski. *Proc. of Int. Symp. on Engineering Geology and Underground Construction*. International Association for Engineering Geology and Environment, pp. 33–43, Springer-Verlag, Heidelberg.

Szechy, K. (1967). *The Art of Tunnelling*. Akademiai Kiado, Budapest.

Taylor, R. N. (1998). Modelling of tunnel behaviour. *Proc. ICE Geotechnical Engineering* **131**, July, pp. 127–132.

Terzaghi, K. (1946). Rock defects and loads on tunnel supports in *Rock Tunnelling with Steel Supports* (eds Proctor, R. V. and White, T.). Commercial Shearing and Stamping Co., Youngstown, Ohio, pp. 15–99.

Van der Berg, J. P. (1999). *Measurement and Prediction of Ground Movements around Three NATM Tunnels*. PhD thesis, University of Surrey.

Ward, W. H., Tedd, P. and Berry, N. S. M. (1983). The Kielder experimental tunnel: final results. *Géotechnique* **33**, No. 3.

Woods, R. I. and Clayton, C. R. I. (1993). *The Application of the CRISP Finite Element Program to Practical Retaining Wall Problems: Retaining Structures*. Thomas Telford, London, pp. 102–111.

7 Settlement

7.1 Prediction of ground movements

Ground movements caused by tunnelling can have a significant impact on overlying or adjacent structures and therefore require consideration when choosing the tunnelling method. This is particularly true when tunnelling in soft ground beneath urban areas. Ground movements due to tunnelling in rock are not usually a problem except where the cover is relatively shallow, in portal areas, or where groundwater may be affected in overlying soils susceptible to settlement.

7.1.1 Characterisation

In cohesionless soils (sands and gravels) the ground movements are usually characterised by an almost immediate movement due to elastic/plastic relaxation and loss of ground in the vicinity of the tunnel shield. In cohesive materials (clays and silts) these relatively rapid movements are followed by a secondary phase of movements due to consolidation, the speed of which depends on the permeability of the material. It is usually the first phase of movement that is of principal concern because the secondary phase may occur over a wider area and consequently cause less angular distortion in nearby structures.

Martos examined the shape of the subsidence trough above mining excavations (1958) and he proposed that it could be well represented by a Gaussian distribution curve. Later, Schmidt (1969) and Peck (1969) showed that the surface settlement trough above tunnels took a similar form.

7.1.2 Models and methods

7.1.2.1 Gaussian model O'Reilly and New (1982) developed the Gaussian model by making the assumptions that the ground loss could be represented by a radial flow of material toward the tunnel and that the trough could be related to the ground conditions through an empirical 'trough width parameter' (K). The model was guided by an analysis of case history data. These assumptions allowed them to develop equations for vertical and horizontal ground movements that were also presented in terms of ground strain, slope and curvature (both at and below the ground surface). The equations have since become widely used particularly during the design process to assess the potential impact of tunnelling works.

The base equations are given as:

$$S_{(y,z)} = S_{(max,z)} \exp(-y^2/2(Kz)^2)$$

$$V_s = (2\pi)^{1/2} Kz S_{(max,z)}$$

$$H_{(y,z)} = S_{(y,z)} y/z$$

where $S_{(y,z)}$ and $H_{(y,z)}$ are the vertical and horizontal components of displacement respectively at the transverse distance y, and the vertical distance z, from the tunnel axis; $S_{(max,z)}$ is the maximum surface settlement (at $y = 0$) and vertical distance z from the tunnel axis; K is an empirical constant related to the ground conditions (e.g. 0.5 for London Clay and 0.25 for some sands and gravels) – note that the product Kz defines the width of the trough and corresponds to the value of y at the point of inflexion of the curve (for most practical purposes the total trough width can be taken to be $6Kz$); V_s is the settlement volume per unit advance.

O'Reilly and New (1982) stated that these equations tend to become inaccurate in the vicinity of the tunnel (within about one tunnel diameter) and Mair *et al.*, 1993 give a method to improve the prediction of sub-surface ground movements for tunnels in London Clay (see also Section 7.1.2.3 below). Their approach is to consider the trough width parameter, K, to be a variable dependent on the depth of the tunnel. This method is effectively the same as that given above except that the loss of ground is considered to occur at a depth of about 1.5 times the tunnel depth (rather than at tunnel axis level) and a different, but constant, value for K.

7.1.2.2 Advance settlement

The equations given in Section 7.1.2.1 describe the form of the ground movements in two dimensions normal to the tunnel axis. In practice the settlement trough also proceeds in advance of the tunnel face. It is a natural consequence of the assumption of a Gaussian transverse profile that this trough should take the form of a cumulative probability distribution and this has been demonstrated by Attewell and Woodman (1982).

7.1.2.3 Three-dimensional models

Tunnelling works often comprise a variety of intersecting excavations where tunnels change in diameter and where cross-connecting adits and other openings occur. New and O'Reilly (1991) incorporated the radial flow and trough width parameter assumptions into the cumulative probability distribution model to provide a three-dimensional model, and demonstrated its application to a relatively complex excavation.

New and Bowers (1994) further developed the cumulative probability distribution model by refining assumptions regarding the location of ground loss and give a full array of equations for the prediction of ground movements in three dimensions. In particular this approach gives significantly improved predictions in the vicinity of the tunnels. This model was validated by extensive field measurements taken during the construction of the Heathrow Express trial tunnel and elsewhere. Also suggested is a method for the prediction of movements caused by shaft sinking.

7.1.2.4 Influence of construction method

Settlement predictions are usually carried out using empirically-based procedures without specific regard to the method of construction. However, the proposed construction method will influence the value taken to represent the volume of the settlement trough and thereby the predicted ground movements. Where ground movements are considered important every effort must be made to control the ground as early and effectively as possible at each stage of the excavation and support process (see also Section 5.3).

The convenience of the Gaussian/cumulative probability distribution curves leads to a series of straightforward mathematical transformations and an apparent precision that may not always be shown in field data. In practice unexpected ground conditions or poor tunnelling practices can lead to significantly larger than predicted ground movements. The considerable strength of the technique lies in its ease of use and in its general validation by field measurements from many sources over many years.

It is of little practical consequence to the ground movements whether the ground loss occurs at the tunnel face or at the periphery of the shield or lining. The construction method will not usually influence the final shape of the ground movement profile but the construction sequence can alter the maximum angular distortions experienced in a direction parallel to the tunnel axis.

7.1.2.5 Numerical modelling Clough and Leca (1989) have reviewed finite element methods applied to the analysis of ground movements around soft ground tunnels. They point out that the soil tunnelling problem has proved resistant to this form of modelling because it is complex, involves parameters which are often poorly defined, and is unforgiving if the analyst does not properly model both the soil and tunnel supports, as well as the construction process. The sensitivity of numerical models to these factors has meant that they are less reliable in predicting 'greenfield' ground movements than the empirically based models. However, numerical models can be of considerable assistance in problems involving soil/structure interactions and the calculation of lining loads.

7.2 Effects of ground movements

7.2.1 Buildings

Given predictions of ground movements it will often be necessary to quantify their potential effects on brick and masonry buildings. Burland (1995) and Mair *et al.*, (1996) have considered this problem. Broadly speaking their approach is to calculate the tensile strains in the building and to interpret these in terms of damage 'degrees of severity', which are expressed in six categories ranging from 'negligible' to 'very severe'. Each category of damage is described and its ease of repair indicated.

For tunnelling works they suggest a three-stage approach:

7.2.1.1 Preliminary assessment Ground surface settlement contours are drawn (using the empirical predictive methods given above) and if the predicted settlement of a building is less than 10 mm it is assumed to have a negligible risk of damage and the assessment process is terminated. This is subject to an additional check that no building experiences a slope in excess of 1:500 (0.2%). (Note that for a given settlement a small shallow tunnel will be more damaging than a large deep one because the structural distortion will be greater for the former.)

7.2.1.2 Second stage assessment The maximum tensile strain in the building is calculated and a 'possible' damage category assigned. Note that this will be a conservative assessment because the building strains are based on 'greenfield' ground movement predictions whereas in practice the actual movements may be reduced by the stiffness of the building. This effect could give rise to problems where services enter buildings because of the differential movement of the building and the adjacent ground.

7.2.1.3 Detailed evaluation This stage is undertaken for buildings predicted to have a 'moderate' level of damage in the second stage. It considers tunnelling sequence, three-dimensional aspects, specific building detail and soil/structure interaction. The assistance provided by numerical methods may be valuable at this stage, and Potts and Addenbrooke (1997) have suggested a hybrid (Gaussian/ finite element) approach. Protective measures would then be considered for buildings remaining in the 'moderate' or higher damage categories.

7.2.2 Pipelines

Bracegirdle *et al.* (1996) have considered the effect of tunnelling on pipelines. They addressed the important problem of cast-iron pipelines, which are particularly vulnerable because of their brittle nature. Again the ground movements are estimated following the equations given by O'Reilly and New (1982) and the tensile strains in the pipe calculated assuming either flexible or rigid jointing of the pipe sections. These strains are compared with various acceptability criteria. Methods of dealing with ground movements that might affect cast-iron pipelines include flexible jointing systems and shorter lengths of pipe to provide some degree of articulation.

7.2.3 Piled structures

Recent difficulties in dealing with the effects of tunnelling on piled structures have been addressed but little, if any, practical guidance is given in the literature, and each case tends to be dealt with on an *ad hoc* basis. Mair and Taylor (1997) have presented two case histories and reviewed published results of numerical models and model tests. However, very few case histories exist and further validation of the models by comparison with field observations is required. The problem of piling close to existing tunnels has been discussed by Chudleigh *et al.* (1999) and is the subject of current research.

7.3 Compensation grouting

Compensation grouting is most commonly carried out in association with new excavation, which may be adjacent to or beneath existing tunnels. The grouting may be intended to protect overlying structures or the existing tunnels. The effects of excavation and compensation grouting will depend largely on the position of each in relation to the existing tunnels and the sequence of grouting and excavation employed. It is generally accepted that there are risks in grouting above an advancing face and therefore an assessment of the distribution of applied pressures and the impact on partially or fully completed linings is required. Grouting either in advance of or following completion of excavation may reduce adverse effects.

7.3.1 Effects on linings

Compensation grouting can, when carried out in close proximity to existing tunnels, induce modes of deformation in linings that are far more damaging than elliptical deformation, which usually accompanies the general loading and unloading of tunnels. When carrying out monitoring of tunnel linings during grouting, therefore, it is not sufficient simply to measure diametric change. It is necessary to determine the mode of deformation and, in the case of bolted segmental cast-iron linings, determine tensile strains at critical locations. Damage to cast-iron linings arising from compensation grouting is usually in the form of tensile cracking of the flanges at bolt positions. Damage in the form of linear cracks along the long axis of the pans of segments may also be seen. Linings where adjacent rings have been 'rolled' are particularly susceptible to damage. Where damage is expected, it may be prudent to release bolts and allow some articulation of the linings.

Damage may also occur where tunnels are lifted by compensation grouting beneath them. In this case, damage may occur as the tunnel linings articulate in the longitudinal direction, and may be concentrated at changes in section, headwalls, etc.

The form of damage to tunnel linings, which can be caused by compensation grouting, can be due to either excessive deflection or excessive stress causing cracking. In segmental linings large deformations can often be accommodated by rotation or shear at the joints between segments without inducing high stresses in the linings themselves. Exceptions include rolled joints, junctions, headwalls or temporary internal propping which substantially increase the stiffness of the tunnel structure. Shotcrete linings do not have the potential to accommodate large movements without cracking.

Compensation grouting is normally carried out in stages that include a pre-consolidation or pre-heave phase to tighten the ground. The process of compensation grouting involves the injection of grout into the ground at high pressure. It is essentially a jacking operation, which produces movement of the ground regardless of whether the grout injected is concurrent with tunnel construction or observationally, afterwards. Thus a reaction force is necessary in order to generate the required (upwards) movement and, consequently, there is unquestionably the potential for loading and deformations to be generated in any tunnel lining or temporary works supports situated below an area of grout injection.

7.3.2 Controlling factors

The controlling factors can be divided into those determined by the design of the grouting facilities and those that relate to the implementation of injections for a given situation.

7.3.2.1 Design of grouting system

- *the vertical total stress at the grouting horizon which has a strong influence on the grout pressure in the ground*
- *the vertical spacing between the grouting horizon and the tunnel or excavation which influences the spread of load*
- *combination with other factors causing loading to a tunnel, for example close proximity tunnels*
- *the properties of the tunnel lining (stiffness, strength, joints).*

The design of a compensation grouting system should include consideration of the potential effect on the tunnel(s) above which it is to be implemented. Observation is a vital part of any such system and measurements of the effect of grouting on tunnels should be included in the monitoring system.

7.3.2.2 Implementation

- *the timing of injections relative to excavation*
- *the volume of individual injections*
- *the plan location of injections relative to the excavation*
- *the properties of the grout particularly with respect to the shape and extent of the grout bulb/fracture formed*
- *the quantities of grouting undertaken.*

If compensation grouting is carried out over a wide area the total vertical stress cannot exceed the overburden pressure. Tunnel linings are generally robust and are able to sustain full overburden pressure. Difficulties therefore arise when the grouting is concentrated in a small area and produces localised loadings or deformations within the lining.

7.3.2.3 Alleviation The implementation of grouting injections can be modified in respect of the controlling factors listed above to reduce the potential impact on tunnels. For example, if compensation grouting is carried out concurrent with the tunnel advance, exclusion zones can be imposed around the excavation face, that is no grout is injected within specified plan distances of the tunnel face. Within limits, the precise location of injections during tunnelling has a negligible effect on the efficiency of the grouting in reducing settlements, while this can be used to reduce the impact of the grouting on the tunnel below.

7.4 References

Attewell, P. B. and Woodman, J. P. (1982). Predicting the dynamics of ground settlement and its derivatives caused by tunnelling in soil. *Ground Engineering* **15**, pp. 13–22.

Bracegirdle, A., Mair, R. J., Nyren, R. J. and Taylor, R. N. (1996). A methodology for evaluating potential damage to cast iron pipes induced by tunnelling. *Proc. Int. Symp. on Geotechnical Aspects of Underground Construction in Soft Ground* (eds Mair, R. J. and Taylor, R. N.). Balkema, Rotterdam, pp. 659–664.

Burland, J. B. (1995). Assessment of risk of damage to buildings due to tunnelling and excavation. *Invited Special Lecture to IS-Tokyo '95, 1st Int. Conf. on Earthquake Geotechnical Engineering*, Balkema, Rotterdam.

Chudleigh, I., Higgins, K. G., St John, H. D., Potts, D. M. and Schroeder, F. C. (1999). Pile tunnel interaction problems. *Conf. on Piling and Tunnelling, London*.

Clough, G. W. and Leca, E. (1989). With focus on use of finite element methods for soft ground tunnelling. Tunnels et Micro Tunnels en Terrain Meuble, B du Chantier à la Theorie. Presses de l'École Nationale des Ponts et Chaussees, Paris, pp. 531–573.

Mair, R. J. and Taylor, R. N. (1997). Bored tunnelling in the urban environment. *Theme lecture to 14th Int. Conf. on Soil Mechanics and Geotechnical Engineering, Hamburg*, Thomas Telford, London.

Mair, R. J., Taylor, R. N. and Bracegirdle. A. (1983). *Géotechnique* **43**(2). Thomas Telford, London, pp. 315–320.

Mair, R. J., Taylor, R. N. and Burland, J. B. (1996). Prediction of ground movements and assessment of building damage due to bored tunnelling. *Proc. Int. Symp. on Geotechnical Aspects of Underground Construction in Soft Ground* (eds Mair, R. J. and Taylor, R. N.). Balkema, Rotterdam, pp. 713–718.

Martos, F. (1958). Concerning an approximate equation of the subsidence trough and its time factors. *Proc. Int. Strata Control Congress, Leipzig*. Deutsche Akademie der Wissenschaften zu Berlin, Section fur Bergbau, pp. 191–205, Berlin.

New, B. M. and Bowers, K. H. (1994). Ground movement model validation at the Heathrow Express trial tunnel. *Proc. Tunnelling '94 Conf.* Chapman and Hall, London, pp. 301–326.

New, B. M. and O'Reilly, M. P. (1991). Tunnelling induced ground movements: Predicting their magnitude and effects. *Invited review paper to 4th Int. Conf. on Ground Movements and Structures, Cardiff*. July, ICE, London.

O'Reilly, M. P. and New, B. M. (1982). Settlements above tunnels in the United Kingdom: Their magnitude and prediction. *Proc. Tunnelling '82 Conf.* IMM, London, pp. 173–181.

Peck, R. B. (1969). Deep excavations and tunnelling in soft ground. *Proc. 7th Int. Cong. on Soil Mechanics and Foundation Engineering, Mexico*. State of the Art volume, pp. 225–290, Institution of Mining and Metallurgy.

Potts, D. M. and Addenbrooke, T. I. (1997). A structure's influence on tunnelling-induced ground movements. *Proc. ICE Geotechnical Engineering* **125**, April, pp. 109–125.

Schmidt, B. (1969). Settlements and ground movements associated with tunnelling in soil. PhD thesis, University of Illinois, Urbana-Champaign, Illinois.

8 Instrumentation and monitoring

8.1 Introduction

Instrumentation is installed typically to:

- *obtain 'baseline' ground characteristics*
- *provide construction control*
- *verify design parameters*
- *measure performance of the lining during, and after, construction*
- *monitor environmental conditions (e.g. settlement, air quality and effects on the groundwater regime)*
- *to carry out research to enhance future design.*

Instrumentation may also be installed to monitor mitigation measures and aid the quantification and management of risk to third parties.

In the context of tunnel lining design, instrumentation and monitoring (I & M) is used to assess the pre-construction ambient state, the geotechnical parameters required for design, for performance monitoring during construction, for design verification, and for long-term post-construction monitoring.

This section deals with the use of instrumentation and monitoring within the design process, for both rock and soft-ground tunnels.

8.2 Value of instrumentation and monitoring

Instrumentation and monitoring will usually be required for a tunnelling project with the degree of effort and expense employed depending upon the nature of the design, perceived hazards and level of risk. It could be argued that, for well-established designs in well-known ground conditions, the necessity for I & M is small. For example, I & M for deep, shield-driven, utility tunnels in stiff clay, supported using standard, precast, reinforced concrete segments is often limited to surface levelling points along and transverse to the line of the tunnel. These points are used to ascertain baseline conditions when assessing claims for damage to third party interests, and to monitor the ground movements due to the tunnel construction process.

However, in high-risk environments, such as shallow tunnels with low cover and variable ground conditions, or where significant ground movements are predicted, consideration of alternative excavation, support and lining designs and a more rigorous approach to the specification of I & M may be called for. For example the *ICE Design and Practice Guide* for sprayed concrete lined (SCL) tunnels (Institution of Civil Engineers, 1996), suggests that where tunnelling induced settlement assessments predict Category 2 damage or greater (BRE Digest 251 – Building Research Establishment, 1990), a complete monitoring system, comprising in-tunnel, borehole and surface instrumentation should be employed.

Similarly, where a wide range of values of design parameters is feasible, I & M to check the response of the structure and the validity of the chosen values may be required. In this context the designers may adopt an observational method such that I & M becomes an integral part of the design verification and construction monitoring process.

Tunnel lining design guide. Thomas Telford, London, 2004

8.3 Existing guidance

There is a large body of published literature providing information on the types of instrumentation available, the suitability of the instrumentation to various tunnelling situations and recommendations on planning and execution of an instrumentation programme. Key texts include those published by the British Geotechnical Society (1973), Cording *et al.* (1975), Hanna (1985), Bieniawski (1984), Dunnicliff and Green (1993) and the Institution of Civil Engineers (1989).

The International Society for Rock Mechanics (ISRM) has also published a number of relevant guides (Brown, 1981) and the BRE has published 'digests' providing guidance on specific instrumentation to monitor the effects of ground movements on existing structures.

A joint industry/UK Department of the Environment, Transport and the Regions (DETR)/UK Engineering and Physical Sciences Research Council (EPSRC) collaborative research project has been undertaken looking at extensive instrumentation data obtained from the Jubilee Line Extension (JLE) project. The research was carried out by Imperial College, London, sponsored by the UK Construction Industry Research and Information Association (CIRIA), London Underground Limited and other industry organisations. Results of the research were published at a conference held at Imperial College in July 2001 by Burland *et al.* (2001). Useful information on the practical aspects of specifying, installing and monitoring instrumentation and case history examples is given.

Due to the large volumes of data generated by the JLE project, the Association of Geotechnical Specialists' (AGS) data management protocol was developed to incorporate mapping, condition survey, and borehole and instrumentation data. Known as 'GEMINI' (GEotechnical Monitoring INformation Interchange) this work has been briefly described by Black *et al.* (2001) and was incorporated as a module (known as AGS-M) within the Version 3 update of the AGS data transfer format in April 2002.

8.4 Instrumentation and monitoring and lining design

8.4.1 General

A brief listing and description of typical instrumentation equipment that may be employed in tunnel monitoring is included for initial guidance in the appended table at the end of this chapter. The quoted ranges, accuracies and precisions are indicative of typical instruments available at the time of writing and may vary with the different manufacturers and with further technical development. The engineer looking to procure such instruments should obtain the latest data for specific instruments from the manufacturers.

Recommendations for the layout and spacing of instrumentation arrays will not be discussed in detail, as this will inevitably be specific to each site. Similarly, the use of I & M for the purposes of back-analysis, research and monitoring environmental effects will not be discussed in any detail. A checklist of the issues to be considered when designing the layout of the I & M arrays is given later in this section (Fig. 8.1).

The layout and spacing between instrumentation arrays will depend on factors such as the stratigraphy, level of detail (volume of data) and degree of redundancy required.

Instrumentation and monitoring is typically employed to provide data in the following areas:

- **greenfield ground response** – earth and water pressures, displacement and strain

- **ground–structure interaction** – relative structural displacements, structural tensile and compressive strains and earth pressures
- **ground–lining interaction** – lining strength and stiffness, relative displacement (distortion), tensile and compressive stresses and strains in the lining and earth pressures
- **ground conditions** – borehole instrumentation to assess ambient earth and pore pressure conditions, groundwater regime and chemistry, face logging for geological conditions, forward probing, groundwater inflow rates, external changes in earth and water pressure
- **monitoring of control and mitigation measures** – (such as compensation grouting or ground freezing): temperature, earth and water pressures, grout–take and pressures, displacement and strain
- **monitoring environmental effects and working environment** – such as ground settlement, noise and vibration, air quality.

Not all of the items listed above will have a direct impact on the design of the lining. The third and fifth points are likely to be most relevant (see Section 5.8 on lining distortion).

Monitoring of tunnel lining behaviour is usually carried out to monitor performance during construction, to assess the development of radial and tangential loads with time (e.g. Barratt *et al.*, 1994) and to monitor lining distortion (convergence) in order to confirm design assumptions. An example of such an application would be instrumentation installed to monitor a tunnel supported using sprayed concrete lining (e.g. Beveridge and Rankin, 1995).

Monitoring may also take the form of strain gauges attached to reinforcement in pre-cast or cast in situ concrete linings or fixed to the flanges of cast-iron segments for example. Loads cells may be used to monitor the performance of arch rib supports.

Monitoring of the lining performance may also be carried out to enhance future design. For example the United States Bureau of Mines (USBM) and British Coal carried out research programmes in the 1960s and 1970s to monitor the performance of shaft linings in the strata of coal measures at considerable depths. Data from these research programmes provided significant insights into the load-deformation interactions and enabled significant design improvements (see Snee in Cutler, 1998).

Although monitoring changes in earth and water pressures due to tunnelling may not have a direct impact on design, a good understanding of the ambient or 'baseline' conditions is essential. It is suggested that the designer have input into planning of the ground investigation whereby boreholes sunk at that stage might also be instrumented for subsequent monitoring of the tunnelling process. Early installation of instruments will improve the designer's confidence in the potential range of design parameters, for example tidal effects in groundwater pressure, which may have a bearing on lining performance and stability.

Monitoring of external control and mitigation measures may include I & M to control ground freezing (see Section 7.3) or compaction and compensation grouting (see Section 3.7.3). For example pressures exerted by compensation grouting may apply permanent or transient loads to an existing lining and induce ground heave. Similarly the design of SCL tunnels in frozen ground will require assessment of the temperature effects upon curing and long-term strength of the concrete.

Tunnel lining design guide. Thomas Telford, London, 2004

Objective	Instrumentation	• Range • Resolution • Accuracy	Comments
Relative vertical movement	BRE-type levelling sockets and precise levelling pins installed on structures, settlement monuments, geodetic surveying targets in structures or tunnel linings	• any • 0.1 mm • 0.5–1.0 mm	Includes tunnel crown levelling points; direct measurement of ground response; can be compared to empirical estimates for rapid assessment; automated theodolites can be employed; surface points may be affected by construction of pavement or road – that is, separations and 'bridging' may occur between pavement and underlying ground. When measuring very small movements, closure errors/accuracy may mask initial trends and vary according to surveyor; surface measurements are an indirect measure of tunnelling performance at depth; time consuming – data frequency limited due to manual operation; coverage may be limited due to access restrictions; levelling in some tunnel environments may achieve realistic accuracy of only 2 mm.
	Precise liquid level settlement gauges with LVDTs installed in surface structures	• 100 mm • 0.01–0.02 mm • ±0.25 mm	Direct measurement of ground/structure response; volume changes due to, say, temperature normally affect all gauges equally and can be eliminated during calculation (however, if one gauge is in a warm tunnel, and another is at the portal, for example, temperature can be a factor); risk of vandalism and effects of exposure to weather; require water and air pipes over significant distances and a stable reference gauge pot.
	Borehole magnet extensometer	• any • ±0.1 mm • ±1 mm–5 mm	Includes high precision magnet extensometer probe; simple and robust, utilises inclinometer casing thereby providing dual function in one borehole; accuracy ±0.2 mm with an electronically controlled motor unit; sub-surface data can be obtained; subject to operator variations; manually operated 'dipper' typically used – time consuming and limiting data frequency.
	Borehole rod or invar tape extensometers	• 100 mm • 0.01 mm • ±0.01 mm–0.05 mm	Direct measurement; simple installation; can measure multiple points in one hole; can be data-logged when using VW/LVDT gauges; can measure both settlement and heave; stainless steel rods may be subject to temperature variations; head requires protection; when logging continuously (i.e. in 'real time') actual data will only be at the frequency that the collar is levelled – that is manually; when using a deep datum it is assumed that no movement occurs – may not be the case; rapid changes may cause temporary loss of VW transducer – dynamic transducer may be required; can also be installed in-tunnel to monitor movements normal to tunnel boundary; accuracies with LVDT: ±10 µε; VW gauge: ±1 µε.
	Satellite geodesy	• Any • to ±50 mm • to ±1 mm	Satellite based levelling techniques include Differential GPS (Global Positioning Satellite) and InSAR (Synthetic Sperture Radar Interferometry). Quality of data can vary with topography, vegetation cover, availability of reflector targets, satellite orbit, and atmospheric effects. Generally applicable to long term monitoring of 'regional' movements at the present time.

Fig. 8.1 Typical applications of instrumentation in tunnelling

Objective	Instrumentation	• Range • Resolution • Accuracy	Comments
Lateral displacement	Surface horizontal BRE invar wire extensometers	• 0.01% • 0.001– 0.05% • 0.01–0.05 mm	Continuous monitoring array possible; direct measure of horizontal strain; require 100 mm diameter telescopic ducting up to 20 m in length to be installed, linked in series between instrument houses; requires substantial installation effort.
Change in inclination	Borehole electrolevels; electrolevel beams on structures and in tunnels; 'tilt meters'	• 50 mm/m (to 175 mm/m) • 0.05 mm/m (to 0.3 mm/m) • to 0.1 mm/m	Data-logged; borehole installations relatively unaffected by temperature variations; additional ground information can be obtained from borehole; can be used to measure longitudinal distortions along tunnels when continuous strings employed; borehole tilt meters and electrolevels can measure tilt in two orthogonal planes; borehole instruments require corrosion protection from groundwater; resolution dependent on beam length. Accuracy can vary with manufacturer.
	Borehole inclinometer probes	• ±53° from vertical • 0.04 mm/m • ±6 mm/25 m	Can be coupled with spider magnet extensometers to obtain the complete movement vector. When interpreting results, can be difficult to pick up small movements.
	Horizontal borehole deflectometer	• ±50 mm • ±0.02 mm • ±0.1 mm	Measures horizontal and vertical deflections. Cannot be used with standard inclinometer casing.
Changes in earth pressure	'Push-in' total pressure cells	• up to 1 MPa • up to 0.1% FS • up to 1.0% FS	Direct measure of changes of pressure in the ground; can be coupled with a piezometer cell to obtain changes in effective stress; can be data-logged using VW transducers; may not be able to obtain actual earth pressures due to installation effects – relative changes only; may require settling-in period of some weeks.
Changes in water pressure	Standpipe piezometers	• any • ±10 mm • ±10–20 mm	Simple to install; robust; rendered ineffective if water table drops below response zone; unable to assess 'real-time' fluctuations in piezometric head due to manual reading and 'lag' in response due to head losses in permeable strata; accuracy depends on operator and condition of 'dip-meter'.
	Pneumatic piezometer (pore pressures are balanced by applied pneumatic pressures)	• 0–20 bar • 0.01 bar • 0.5% FS ± 0.02 bar	Analogue, 'membrane switch' (hydraulic transducer) or digital readout can be used; not affected by very low temperatures; may be pushed into soft soils – minimising disturbance; not effective where suctions occur over sustained periods.
	Vibrating wire piezometer	• up to 35 bar, • 0.025% FS • ±0.1% FS	Can be read using a hand-held digital transducer unit, or remotely using a data-logger; standard sensors can measure suctions up to cavitation (suctions up to −1500 kPa can be measured at shallow depth using the Imperial College Suction Probe); instability in readings may occur for rapidly fluctuating piezometric levels; sensors may require settling-in period of some weeks.

Fig. 8.1 (continued)

Tunnel lining design guide. Thomas Telford, London, 2004

Objective	Instrumentation	• Range • Resolution • Accuracy	Comments
Crack or joint movement	Tell-tales	• ±20 mm • 0.5 mm • ±1 mm	Direct measurement of ongoing movement; local point measurement; does not give quantitative measurements of stress and strain; some instruments subject to temperature corrections.
	Calliper pins/micrometer (DEMEC gauges)	• up to 150 mm • 0.02 mm • ±0.02 mm	DEMEC gauge has a more limited range but resolution to 0.001 mm and accuracy to 0.005 mm. Pins simple and inexpensive to install.
	Vibrating wire joint-meters	• up to 100 mm • up to 0.02% FS • up to 0.15% FS	Can measure three orthogonal directions with triaxial device; built-in temperature correction; can be data-logged; simple surface installation but needs to be protected from vandalism.
Strain in structural member or lining	VW strain gauges	• up to 3000 $\mu\varepsilon$ • 0.5–1.0 $\mu\varepsilon$ • ±1–4 $\mu\varepsilon$	High accuracy; direct measurement at a point; generally robust and reliable; can be waterproofed for exposed conditions; gauges can be directly installed on rebar or flanges of cast-iron segments, or on rock bolts; provide information on that member only – no indication of overall structure performance; small gauge lengths result in highly localised measurements; may be susceptible to corrosion or damage if not adequately protected; temperature corrections may be required; pattern of strain may be highly variable and difficult to convert into stress; results may be affected by heat of hydration in concrete during curing, cracking and grouting.
	Fibre optics	• to 10,000 $\mu\varepsilon$ (1% strain) • 5 $\mu\varepsilon$ • 20 $\mu\varepsilon$	Glass cables are light and corrosion resistant; easy to splice cables for long lengths (range from 10 cm to 1 km); can insert many sensor locations along cable length (depending on wavelength of light); can multiplex up to +100 cables; can be embedded in concrete or mounted on a structure; can operate in temperatures between −20 °C and +50 °C.
Tunnel lining diametrical distortion	Tape extensometers across fixed chords	• up to 30 m • 0.001–0.05 mm • ±0.003–0.5 mm	Traditional approach, results 'understood'; simple and portable; direct measurement of relative distortions (only); measurement may disrupt excavation cycle; accuracy may decrease with increasing span; access difficulties may arise in large excavations or shafts; possible interference in construction cycle; results affected by operator experience and temperature fluctuations; cannot be automated; indirect measure of tunnel lining performance.
	3D geodetic optical levelling ('retro' or 'bioflex') targets, levelling diodes or prisms	• any • 0.1–1.0 mm • 0.5–2.0 mm	Rapid monitoring of a large number of points possible; reading can be fully automated and data-logged using motorised instruments; absolute measurements of position obtained; mounting bolts can be used for other measurements such as tape extensometers; in the tunnel environment, usually best to have targets within 100 m of station; monitoring may obstruct construction cycle; indirect measure of tunnel lining performance; probably the most common method used to monitor distortion during construction, at the time of writing.

Fig. 8.1 (continued)

Objective	Instrumentation	• Range • Resolution • Accuracy	Comments
Tunnel lining diametrical distortion (cont'd)	Strain gauged borehole extensometers installed from within tunnel	• 100 mm (3000 $\mu\varepsilon$) • 0.01 mm (0.5 $\mu\varepsilon$) • ±0.01–0.05 mm (±1–10 $\mu\varepsilon$)	Direct measurement; simple installation; measure multiple points in one hole; can be data-logged when using VW gauges; accuracy LVDT: ±10 $\mu\varepsilon$; micrometer: ±0.01 mm; stainless steel rods may be subject to temperature variations; head requires protection; the deepest anchor is assumed to be beyond the disturbed zone of influence – if not, relative movements may be underestimated.
	Basset Convergence system	• ±50 mm • 0.02 mm • ±0.05 mm	Interlinked tilt sensor array; permits real-time monitoring/data-logging of lining distortion.
Lining stresses	Total pressure (or 'stress') cells	• 2–20 MPa • 0.025–0.25% FS • 0.1%–2.0% FS	Direct measure of subsequent changes in earth pressure at a point; total pressure (or 'stress') cells installed between lining and ground (tangential pressure cells) or cast into lining (radial pressure cells) utilising membrane switch (read using an oil pressure gauge) or VW transducers. Comprise either mercury (high pressure) or oil-filled (low pressure) cells; can be installed between segment joints; better accuracy and resolution obtained from lower range cells; actual pressures not measured due to relative stiffness effects; installation may affect quality of results – requires experience; primary stress state has already been altered by the excavation; may not give realistic estimates due to localised point loads etc.; often need re-pressurising after lining concrete has cured due to concrete shrinkage; a knowledge of concrete creep and deformation characteristics required during interpretation; post construction testing such as the flat-jack also possible.
Lining leakage	Flow meter	• any • 1 litre/min • 2 litre/min	Indirect measure of overall inflow; simple apparatus; can be data-logged using a submersible pressure transducer.
Vibration	Triaxial vibration monitor/seismograph	• 250 mm/sec • 0.01–0.1 mm/sec • 3% at 15 Hz	Measures PPV and accelerations in three orthogonal axes; portable equipment.

Notes:
1. Quoted range/resolution and accuracy derived from published and trade literature as an indication of relative performance only. May change with ongoing technical development by manufacturers.
2. For borehole installations, additional information can be obtained from logging/in situ testing.
3. Definitions: range = maximum and minimum recordable values for the instrument, resolution = the smallest change that can be recorded by the instrument, accuracy = difference between recorded value and the 'actual' value as quoted by the manufacturers, rather than a measure of field performance; FS = full scale.

Fig. 8.1 (continued)

8.4.2 Observational Method

Instrumentation installed for the purpose of monitoring the performance of the tunnel lining during construction may form a part of the observational method of design (see Section 5.5). The Observational Method (OM) is an example of the close integration of I & M and lining design, and Nicholson *et al.* carried out a comprehensive review of this approach (1999). They defined the method as:

A continuous, managed, integrated process of design, construction control, monitoring and review that enables previously defined modifications to be incorporated during or after construction . . . .

For example instruments installed for NATM or sprayed concrete lined (SCL) tunnels designed according to OM principles is an example of an *ab initio* (i.e. from inception) application. Here instrumentation installed with the primary lining is intended to monitor subsequent interaction between the lining and the ground, and highlight differences in magnitude and trend of results from past records. Where such data indicate substantial divergence, pre-determined contingency measures are put in place.

Alternatively, the OM may be applied as a 'best way out' approach to an unforeseen change in ground conditions or instability of the excavation (Peck, 1969). Peck stated that 'most probable' and 'most unfavourable' ground conditions should be defined when applying the OM. Nicholson *et al.* (1999) recommend that most unfavourable ground conditions may be obtained by applying partial factors of safety to characteristic values according to Eurocode 7, which may then be used to obtain Ultimate Limit State (ULS) predictions. Serviceability Limit State (SLS) predictions may be obtained from most probable and characteristic conditions.

8.4.3 Design checklist

The following list of factors to be considered by the designer is based on a similar list produced by Dunnicliff and Green (1993). Detailed discussion on each of these points will be found in the literature and the list given here is intended as a brief *aid-memoire*.

- **Ground appreciation** Assess stratigraphy, strength, stiffness, in situ stress, compressibility and permeability of the ground, and anticipated magnitude of changes. Any boreholes formed for the purposes of installing instrumentation should be used to enhance the understanding of the ground model.
- **'Greenfield' check** Where possible, install I & M in areas to monitor the 'greenfield' response. This will facilitate the interpretation of the degree of interaction observed in other developed areas of the project.
- **Assess instrumentation limitations** Understand the limitations of what can be measured; for example is there a 'scale effect' associated with different instrumentation? (e.g. measurement of lining stress reflects local response while lining distortion is a global effect). Identify critical areas where additional 'local' instrumentation may be required to get meaningful results.
- **Resolution of values** Establish the required resolution of the instruments based on their likely minimum values, and the required range based on the maximum value expected. However, it should be recognised that instruments with a large range often have a lower resolution than instruments with a small range. For some instruments installation effects may induce a reduction in sensitivity of the instrument, influencing the choice of a suitable range.
- **Necessary instruments only** Identify clear objectives for the instrumentation and install only sufficient instruments that are

necessary to the problem while allowing for redundancy in critical areas. Allow for a reserve of key instruments to be stored on site in the event that unexpected movements occur.

- **Reliability of chosen instruments** Consider durability, reliability, consistency and maintenance and calibration requirements of the chosen instrumentation. Are the preferred instruments tried and tested? Account for possible damage during construction and potential vandalism when planning the instrumentation layout.

- **Who is responsible?** In contract documentation, define clear responsibilities for installation and commissioning, calibration, provision of baseline data within the chosen contractual framework and responsibilities for ongoing maintenance, monitoring, interpretation and reporting (post-contract if required). Ensure, as far as possible, that the monitoring teams and construction teams liase closely to communicate the effects of construction upon the anticipated movements. This may be achieved, for example, by regular site monitoring meetings.

- **Methodology** Ensure adequately trained installation and monitoring personnel are available and method statements are agreed that cover sensor calibration and installation methods. Ensure that all relevant data during site calibration are recorded, for example ambient temperature and barometric pressure.

- **Frequency of readings** Determine frequency of readings required and, where relevant, allow sufficient time before construction commences to obtain essential baseline data. Ensure that interpretation of results, carried out by qualified personnel, accounts for environmental effects (e.g. diurnal, tidal or seasonal variations in pore pressure, atmospheric pressure or temperature) as appropriate. Consider varying reading frequency according to initial results in order to minimise the potential for information overload.

- **Data management** Establish acceptable trigger levels and associated courses of action, the method of data collection, interpretation and presentation (see above). Specifications should not demand excessive data collection but rather permit flexibility to respond to unexpected changes with an increase in reading frequency at specific instruments or arrays. Reporting should be in a readily digestible form to permit timely assessment of key trends and avoid excessive data handling and transmission. There should be a clear hierarchy for passing the data and interpretations to the correct people who have responsibility for any necessary review and consequent action.

- **Real-time monitoring** Assess the requirement for 'real-time' data acquisition on the basis of the anticipated rate of change of the parameters being measured and the requirement for a rapid response. Care must be taken to ensure that the chosen sensor can provide a dynamic response if changes are likely to be particularly rapid. Automated/data-logged systems will incur additional costs in communication systems. The targets should be rugged and placed where least likely to get damaged. In the event of damage they should be replaced immediately and calibrated.

- **Tunnelling data** Wherever possible include in-tunnel logging (face records or TBM data logging) within the monitoring regime.

- **Data format** Consider the use of the AGS format for storage and reporting of data where the scale of the project and volumes of data justify such an approach. The relative merits of the use of software, or more 'labour-intensive' methods for manipulation and analysis of the data should be assessed.

8.5 Management of third-party issues

While not necessarily having a direct impact on design, the management of third-party issues is an important area where I & M is required. The prediction and control of tunnelling induced settlement is one of the principal areas of significant concern when assessing the risks to nearby utilities and structures during initial design. This hazard is usually assessed in a phased manner as discussed in Chapter 7 of this Guide. Modifications to the excavation sequences and, potentially, the tunnelling method chosen in order to mitigate these risks (for example SCL versus TBM excavation) may have a substantial influence on the selection and design of the lining.

When designing the layout of the I & M arrays for tunnels in developed areas it is recommended that, where possible, instrumentation is installed to assess 'greenfield' behaviour if such a location is available. This will provide useful indirect data on the performance of the tunnelling process and will facilitate an assessment of the degree of ground-structure interaction in other areas.

Monitoring of significant existing defects, noted during pre-contract condition surveys in services and structures within the anticipated area of influence of tunnelling induced movement, is also required. Pre-construction condition surveys, agreed to by all parties, are essential *before* work commences on site. A descriptive scheme that is commonly adopted for describing damage to buildings is that proposed by Burland *et al.* (1977).

The other principal environmental effects, excluding changes in pressure and displacement in the ground, are noise, dust and vibration. Pre-construction surveys will be required to establish existing ambient conditions in terms of noise and vibration.

Ground borne vibration, which may also be manifested in a building as re-radiated noise, can be a major constraint and has been an area of considerable recent study (e.g. Hiller and Bowers, 1997). Vibrations may arise both during construction and during operation of the completed tunnel and consideration may have to be given to mitigation of this effect during the preliminary selection of tunnelling method.

Monitoring of in-tunnel conditions (e.g. heat, humidity, air pressure, dust and gas concentrations, and exposure to Hand Arm Vibration Syndrome) should also be considered, as the results of the risk assessment may suggest that a particular tunnelling method and hence lining type are undesirable on health and safety grounds.

In the post-construction period there may be a requirement for long-term monitoring, for example on rail and road tunnels where public health and safety is dependent on the performance of the civil structures. Interpreting the data from monitoring on a continuous real-time basis is open to a number of concerns including:

- *the long-term stability of the instruments*
- *variations in load over time due to creep effects in concrete*
- *time-dependent behaviour of the ground as readjustments of loads in the ground occur with time, due to either creep effects or release of strains resulting from one-off events such as an earthquake (Douglas and Alexander, 1989; Williams et al., 2001).*

Alternatively, simpler instruments can be used with periodical brief possessions for inspections and data collection.

It is necessary to be very clear about the objectives of the long-term monitoring and the instrumentation necessary to achieve those objectives with a sufficient level of confidence.

8.6 Data acquisition and management

8.6.1 General

Data may be obtained from an I & M programme as either manually recorded readings (e.g. precise levelling, dial gauge extensometer readings, manual piezometer readings, etc.) or as digital data (e.g. automated geodetic surveying data, vibrating wire or LVDT transducer output). Advantages with manual reading of instrumentation include reduced risk of data overload, lower installation costs compared to remotely read instruments and additional visual information obtained at the time of reading. Disadvantages include the potential for reading errors and the relatively lower frequency of data points obtained.

Among the advantages of using electronic transducers in conjunction with a data-logging facility (termed 'automatic systems' by Wardle and Price, 1998) are that real-time data and more detailed records (i.e. higher frequency of readings) can be obtained. This can be particularly valuable in research applications, and for the control of compensation grouting (see Section 7.3).

Data-loggers can also be programmed to adjust reading frequency in response to pre-set trigger values and to initiate alarms. A recent application of data-logging facilities on the Channel Tunnel Rail Link (CTRL) North Downs Tunnel project utilised a system whereby mobile telephones were notified with text messages of actual readings when trigger values were exceeded. Disadvantages with such an approach include higher initial installation costs and the potential to lose all data if the logger is damaged or fails.

Transducers are usually individually cabled direct to the data-logger/controller via a 'multiplexer' unit, which samples them at pre-determined intervals and stores the data. Consideration must be given to access to the data-logger when downloading information to a computer. This may be achieved manually, by using a hand-held interrogator unit, and a standard RS-232 connection and software. Alternatively a hard-wired connection to a computer, a telephone connection via a modem or a radio transmitter can be used to enable remote access direct from the office personal computer.

Due to recent improvements in the accuracy of measurement, automated total stations are becoming an increasingly popular approach to monitoring distortions in existing structures and tunnel linings. These comprise high precision programmable total stations, which take electronic distance measurements (EDMs) and angular measurements relative to stable datum points, at pre-determined frequencies, to determine the three-dimensional displacement of reflector targets over time. This approach is utilised for monitoring projects, extensive in area, where the costs of using manual or data-logger based approaches alone is prohibitive.

A number of proprietary software packages are available to interrogate and manipulate the data for presentation and analysis. Care must be taken, however, to ensure that adequate resources are provided to ensure timely interpretation and response to real-time data.

Graphical output is generally preferred. Where, for example, tidal influences are likely to be felt, comparison of the tidal variation

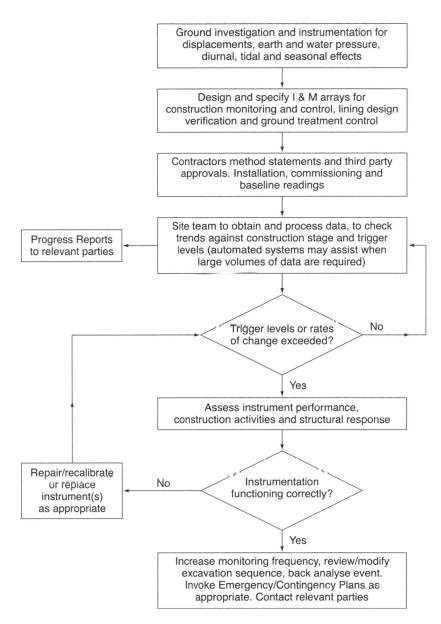

Fig. 8.2 Monitoring of the construction process – simplified flow chart

against the recorded data is useful. Additional information that should be recorded includes environmental conditions (humidity, rainfall, temperature, etc.), excavation progress (position relative to the instrumentation array) and any extraneous events that may have affected the results.

Where the Observational Method has been employed the role of the designer in responding to the results of the I & M system is clear. However, in more traditional contracts the situation may arise where the site supervision team acts independently of the designer. It is recommended that, where possible, the original designer's involvement be maintained so that key assumptions do not remain untested and feedback during construction will benefit future design. In addition insights that the designer may have gained during the design process should be communicated to the construction team and tested against the monitoring data. The designer is best equipped to consider appropriate reactions to an unexpected result or anomaly. Figure 8.2 provides a simplified flow chart illustrating the monitoring of the construction process.

8.6.2 Trigger values

It is normal practice to establish 'trigger values' for key indicator parameters (such as displacement, strain or pressure), which determine appropriate actions in response to these values being exceeded. Typical definitions of these trigger values are as follows.

- **Warning/amber** A pre-determined value or rate of change of a parameter that is considered to indicate a potential problem, but not of sufficient severity to require cessation of the works. For example this level may equate to movements above which a chosen BRE Category of Damage is predicted to occur to buildings and utilities or where routine maintenance thresholds for railway track are just exceeded. Exceeding this trigger level will generally require a check on instrument function, visual inspection of the structure being monitored, increase in monitoring frequency, review of the design and modification of the construction process.

- **Action/red** This level may equate to movements above which an unacceptable BRE Category of damage is predicted to occur to buildings and utilities or where safe operating thresholds for railway track geometry are exceeded. If this value is surpassed an immediate check on instrument function and visual inspection of the structure being monitored will be required, as well as the initiation of a pre-determined response, which may include temporary cessation of work, back analysis of the event and modification of the design and construction process.

It is also recommended that the monitoring records be examined by someone expert in monitoring matters on a regular basis, to ensure that any untoward trends (pre-trigger) are identified and acted upon in a timely fashion.

The selection of appropriate trigger values will depend on the particular requirements of the project and the governing 'failure mechanism(s)' assessed by the designers. For example, Wareham *et al.* (1997) describe the installed instrumentation and selection of trigger levels for a sub-aqueous cast-iron lined tunnel in central London. In that case a governing influence on lining behaviour during the works was identified to be tidal variation in water pressure. Hence, following an initial 'baseline' monitoring period prior to the start of works the trigger levels were determined as a multiple of the maximum distortion and extreme fibre tensile strain observed during tidal cycles, checked against permissible strains in the cast-iron circumferential flanges.

An alternative approach was adopted for the construction of the CTRL North Downs Tunnel where primary support was provided by cement grouted rockbolts, mesh, arch girders and sprayed concrete (see Chapter 10). Trigger levels were determined as the permissible early age (<10-day) strain in the shotcrete lining divided by an appropriate factor of safety for each trigger level (Watson *et al.*, 1999).

Even if serious anomalies are not indicated, it is always worth comparing predictions with observed values in order to understand the behaviour of the structure and ground.

8.7 Case histories

Dunnicliff and Green (1993) refer to the US Federal Highway Administration (FHWA) proceedings (Federal Highway Administration, 1980), which provide a key-worded index of 300 publications on examples of the use of I & M for underground excavations. They reproduce a selection of 25 references from that list. Similarly Hanna (1985) provides a selected list of 19 case histories of tunnel instrumentation. More recent case history data for particular projects may be found in the proceedings of numerous symposia on the subject such as the recent Ninth Géotechnique Symposium in Print (Institution of Civil Engineers, 1994) and CIRIA Special Publication 200 (Burland *et al.*, 2001). A case history for the CTRL North Downs Tunnel is reproduced in Chapter 10.

8.8 References

Association of Geotechnical Specialists (Format for electronic transfer of geotechnical data) – 3rd edition. Downloadable report from www.ags.org.uk.

Barratt, D. A., O'Reilly, M. P. and Temporal, J. (1994). Long-term measurements of loads on tunnel linings in overconsolidated clay. *Tunnelling '94 Conf.* Chapman and Hall, London.

Beveridge, J. P. and Rankin, W. J. (1995). Role of engineering geology in NATM construction. In *Engineering Geology of Construction* (eds Eddleston, M. *et al.*). Geological Society Special Publication No. 10, pp. 255–268, Geological Society, London.

Bieniawski, Z. T. (1984). *Rock Mechanics Design in Mining and Tunnelling*, Chapter 7. Balkema, Rotterdam.

Black, M. G., Withers, A. D. and Pontin, S. (2001). Chapter 19 – Data handling and storage in *Building Response to Tunnelling – Case Studies from the Jubilee Line Extension, London. Volume 1: Projects and Methods* (eds Burland, J. B., Standing, J. R. and Jardine F.). Thomas Telford, London. CIRIA Special Publication 200.

British Geotechnical Society (1973). *Proc. Symp. Field Instrumentation in Geotechnical Engineering*. Butterworth, Oxford.

British Standards Institution (1999). BS 5930 *Code of Practice for Site Investigations*. BSI, London.

Brown, E. T. (ed.) (1981). *Rock Characterization Testing and Monitoring – ISRM Suggested Methods*. Pergamon Press, Oxford.

Building Research Establishment (1990). *Assessment of Damage in Low-rise Buildings, with Particular Reference to Progressive Foundation Movement*. BRE, Garston, Watford. Digest 251.

Burland, J. B., Broms, B. B. and de Mello, V. F. B. (1977). Behaviour of foundations and structures. *Paper presented at the 9th Intl. Conf. SMFE, Tokyo*, July, Session 2. BRE, Watford, CP 51/78.

Burland, J. B., Standing, J. R. and Jardine, F. (eds) (2001). Building response to tunnelling – Case studies from the Jubilee Line Extension, London. *CIRIA Special Publication 200. Volume 1: Projects and Methods*. Thomas Telford, London.

Cording, E. J., Hendron, A. J., Hansmire, W. H., Mahar, J. W., MacPherson, H. H., Jones, R. A. and O'Rourke, T. D. (1975). *Methods for Geotechnical Observations and Instrumentation in Tunnelling. Vols 1 and 2*. Dept Civ. Eng. University of Illinois, Urbana-Champaign, Illinois.

Cutler, J. (1998). Deep trouble. Report on the lecture by Dr C. Snee for the joint BGS/ICE Ground Board informal discussion: 'Geotechnical aspects of shaft sinking' held at the ICE, 8th April, 1998. *Ground Engineering* November.

Douglas, T. H. and Alexander, R. J. (1989). *Dinorwig Power Station: Results of Longer Term Monitoring Instrumentation in Geotechnical Engineering*, Apr. Nottingham.

Douglas, T. H., Arthur, L. J. and Wilson, W. (1984). *Dinorwig Power Station: Observations Related to Monitoring*. International Society for Rock Mechanics, Cambridge.

Dunnicliff, J. and Green, G. E. (1993). *Geotechnical Instrumentation for Monitoring Field Performance*. John Wiley and Sons, London and New York.

Federal Highway Administration (1980). *Proc. Conf. on Tunnel Instrumentation – Benefits and Implementation* (eds Hampton, Browne and Greenfield). FHWA, USA. Report No. FHWA-TS-81–201, Washington, DC.

Hanna, T. H. (1985). *Field Instrumentation in Geotechnical Engineering*. Trans Tech Publications, Uetikon, Zürich.

Highways Agency. *Advice Note on the Application of the New Austrian Tunnelling Method to the Design and Construction of Road Tunnels*. Advice Note BA 95, HMSO, London.

Highways Agency. *Application of the Observational Method in the Construction of Road Tunnels*. Departmental Draft Standard BA 71, HMSO, London.

Hiller, D. M. and Bowers, K. H. (1997). Groundborne vibration from mechanized tunnelling works. *Proc. Tunnelling '97 Conf., London.* IMM, London

Institution of Civil Engineers (1989). Geotechnical instrumentation in practice – purpose, performance and practise. *Proc. Conf. Geotechnical Instrumentation in Civil Engineering Projects.* Thomas Telford, London.

Institution of Civil Engineers (1996). *ICE Design and Practice Guides – Sprayed Concrete Linings (NATM) for Tunnels in Soft Ground.* Thomas Telford, London.

Nicholson, D., Tse, C., Penny, C., O'Hana, S. and Dimmock, R. (1999). *The Observational Method in Ground Engineering: Principals and Applications.* CIRIA, London, Report 185.

Peck, R. B. (1969). 9th Rankine Lecture: Advantages and limitations of the Observational Method in applied soil mechanics. *Géotechnique* **19**, No. 2. Thomas Telford, London, pp. 171–187.

Standing, J. R., Withers, A. D. and Nyren, R. J. (2001). Measuring techniques and their accuracy in *Building Response to Tunnelling. Case Studies from the Jubilee Line Extension* (eds Burland, J. B. *et al.*). Chapter 18. CIRIA, London Special Publication 200.

The observational method in geotechnical engineering (1994). *Ninth Géotechnique Symposium* in Print **XLIV**, No. 4, December.

Wardle, I. F. and Price, G. (1998). Automatic tunnel monitoring. *Tunnels & Tunnelling International*, May, London.

Wareham, B. F., Macklin, S. and Clark, D. (1997). Strengthening the Northern Line: A unique project. Design, construction and tunnel response. In *Proc. Conf. Tunnelling '97*, September. IMM, London.

Watson, P. C., Warren, C. D., Eddie, C. and Jäger, J. (1999). CTRL North Downs Tunnel. *Proc. Tunnel Construction and Piling '99*, September. Brintex/Hemming Group, London.

Williams, O. P., Douglas, T. H. and Graham, J. R. (2001). Dinorwig Power Station: Monitoring and maintenance 1980–2000. *Hydro 2001 Conf., Riva del Garda, Italy*, September.

9 Quality management

9.1 Introduction

The objective of this chapter is to define the main issues to be considered in managing the quality of the design for a tunnel lining, that is ensuring that the required lining is correctly included within the design and that the manufacture and installation of the lining is carried out in accordance with the design. Reference will only be made, therefore, to matters of quality assurance and quality control where they impact directly on the design process. Issues relating to manufacture and construction workmanship, for example, will not be covered unless they detract from the performance of the design. Notwithstanding this qualification, it will be seen how many issues outside the detail design process do impact on the requirements or intentions of the design. It is paramount, therefore, that all information pertinent to the design, for example design parameters, loadings, tolerances, etc., are recorded and available to all parties involved in the design, manufacture and construction of a tunnel lining. The recording of this information will be referred to as the Quality Plan (QP). Design offices with ISO 9001 and ISO 9002 accreditation will have internal procedures that cover much of the following text.

9.2 Design stage

This section applies to the design of all lining types. Separate sections will follow to deal with the quality management issues specifically associated with manufactured and cast in situ or sprayed concrete linings.

9.2.1 Quality Plan

At the commencement of the design process the first Quality Plan should be drafted. This is a live document and may change during the design process. The importance of the document is to set the initial requirements, assumptions, parameters, etc., record any subsequent changes and include reference as to why all subsequent changes have been made.

The first draft will take from the client's brief the specific requirements to ensure that the lining is designed to suit the needs. Examples of such requirements could be:

- *diameter*
- *location/alignment*/depth**
- *design life*
- *use/content*
- *watertightness*.

The importance at this stage is to ensure that the client's needs will be fulfilled by the design without cause to redesign to correct mistaken or omitted requirements.

Having identified the client's requirements, the next stage is to identify those parties, and their roles, responsibilities and duties, that are contributing to the design. This may include both internal parties within the design office/company and all external bodies contributing information to the design process.

---

* May not be specified by the client but derived from other aspects of the design.

Examples of such parties or bodies are:

- *geotechnical engineer*
- *alignment designer*
- *internal fit out designer*
- *client representative*
- *contractor**
- *manufacturer**
- *project manager*
- *engineering manager*
- *designer†*.

The formal standards to which the design is to be undertaken are documented next. These will include the various codes of practice, national or European standards, internal design standards and possibly the client's own design standards. These will relate to all stages and aspects of the design, manufacture, construction and operation of the tunnel in as much as they impact on the design. This will ensure a consistent and co-ordinated approach by all parties involved during the development and life of the lining. Subsequent sections will deal with how recorded data at this stage is transferred to other parties and used during future stages of the lining's development.

At this stage the quality plan can start to address and document the specific design considerations more commonly associated with a lining design. Examples of these are:

- *ground conditions*
- *soil parameters*
- *groundwater levels*
- *groundwater quality*
- *surcharge loadings (present and future, constant and cyclic)*
- *internal loadings and/or pressures*
- *construction loadings, for example TBM thrust, back grouting, etc.*
- *imposed loads from geotechnical processes, for example compensation grouting*
- *handling and transportation induced loading*
- *maximum segment piece weight*
- *gasket provisions*
- *material‡*
- *segmental or in situ.‡*

This is not intended to be an exhaustive list but an example of how all considerations included in the design should be made visible for all to see. Primarily this will be for the design checker or verifier but in later stages the subsequent manufacturers, constructors and users of the lining will need to be able to refer to this information.

The remaining information to be documented before design commences is the methods of analysis to be used. This will include all software packages with sufficient detail to identify the principles of the analysis.

In all cases it may be found that factual data cannot be ascertained but is relevant to the design. In such cases assumptions can be made providing they are documented within the plan. The

---

* If known at this stage.
† The person/party responsible for co-ordinating all involvement and producing the final design.
‡ If known or defined at this stage.

Tunnel lining design guide. Thomas Telford, London, 2004

validity of assumptions will be subject to the approval of the clients where they fall within their domains, or by the checker or verifier of the design.

The checking and verification process should be clearly defined within the Quality Plan (QP).

The recipients of the QP should be clearly identified together with the revision number of the current issue.

9.2.2 Design development statements

During the designing of a lining additional situations or information may arise, possibly as a result of the design, which necessitate changes to the documented details within the QP. Such changes are acceptable providing they are documented and include reference to their implications and why they have been made.

The process of documentation is most visible if recorded as a separate document known as a design development statement. The content of the statement, once signed off as valid under the checking/verification process, should be incorporated within the QP as an addendum or a redraft and circulated to all named recipients.

9.2.3 Design outputs

The outputs of the design process will essentially be the calculations, drawings and specifications including certification of checking and verification of the design as required.

The output documents will be required by either a manufacturer or contractor, depending on the type of lining designed, that is precast/manufactured or cast in situ.

The production of the lining must be driven from the approved output documents, which may include, but are not limited to, the examples given above. While there should be no duplication of information in an unverified form it is prudent to document the principle outputs and refer to the verified output documents as a 'Final Design Statement' so as to guide the recipient of the design in producing the lining.

9.3 Manufactured linings

This section will deal with specific quality management issues relating to the manufacturing process of factory produced linings. While this essentially refers to segmental concrete linings the principles could apply to any other material types.

9.3.1 Quality Plan

Manufacturers will receive the design output documents and design statement as described above. If they are to carry out any element of design they should also receive the final Design Quality Plan to ensure they comply with the prescribed standards and requirements of the design plan.

On receipt of this information the manufacturers will develop their own QPs to ensure and demonstrate compliance with the design. Companies accredited to ISO 9001 and ISO 9002 will have standard procedures to draft such a plan.

While this is not the responsibility of the designer he or she may wish to see the manufacturer's QP and be aware that it demonstrates that:

- *dimensions and tolerances of manufacture are correct*
- *reinforcement and cover is incorporated correctly*
- *fixing details, grout holes/plugs, lifting arrangements are correct*

- *materials are of the correct strengths, properties and standards*
- *required surface finishes will be achieved*
- *correct quality control procedures are in place for material sources and the manufacturing process.*

9.3.2 Quality control

A company with ISO 9001 and ISO 9002 accreditation will have standard procedures to monitor and control the quality of the finished product. The procedures will commence with checks on quality compliance at the source of materials and continue throughout the manufacturing process to ensure the deliverables of the QP are achieved.

Quality control (QC) procedures are not so related to the design process that they warrant further discussion within this document. However, designers should be aware that the manufacturer's QC procedures should be sufficiently frequent to minimise the risk of non-conformities arising in products dispatched from the factory.

Two specific QC tests that are of more interest to the designer are the proof ring build and the load test. The former will be conducted at the outset of manufacture where two rings are built in the factory to demonstrate compliance with dimensions and tolerances. Main production runs will not usually commence until compliance has been demonstrated by this test. The test should be repeated after a prescribed number of uses of the production moulds to demonstrate continued compliance. The designer should specify the frequency of the tests and the British Tunnelling Society's *Specification for Tunnelling* gives guidance on this matter.

Test loading of rings to demonstrate the adequacy of the design and manufacture of the lining is possible although it is a more complex issue. Defining such tests is outside the scope of this section. As a quality management tool, load tests should be carefully designed to prove acceptable axial, flexural and combined loading cases.

9.3.3 Manufacture outputs

The manufactured lining should leave the factory with proof that it has been manufactured in accordance with the design. Documentation (including relevant test certificates) will adequately record compliance.

In addition, the documentation should include detailed handling and build requirements that allow the lining to be installed without exceeding any loading conditions included within the design. Such requirement could include but need not be limited to:

- *ring erection details*
- *maximum TBM thrust*
- *maximum torque on fixings*
- *maximum grouting pressures*

9.4 Cast in situ and sprayed concrete linings

This section will deal with the specific quality management (QM) issues relating to cast in situ and sprayed concrete linings in so much as they relate to the design process. While the manufacture and installation are concurrent operations carried out on site, there are similarities in controlling quality to factory manufactured linings.

9.4.1 Site quality plan

In common with manufactured linings, this will prescribe how the deliverables of the design are achieved during construction. The plan will be drafted from the Design Quality Plan and expanded to include issues such as:

- *excavation sequences*
- *lining thicknesses*
- *temporary supports to be incorporated into the lining*
- *waterproofing membranes*
- *concrete strengths*
- *reinforcement (rebar) details*
- *material types and sources*
- *batching and mixing requirements*
- *construction processes*
- *quality control procedures*
- *monitoring.*

The purpose of this section is not to describe what should be done under each heading but to illustrate that the site QP includes sufficient detail to ensure that the lining is built in accordance with the design and that adequate QC procedures exist to maintain the desired quality of construction.

9.4.2 Site quality control

This is a specific site-based operation and the QC procedure should be drafted accordingly. However, the designer may wish to specify particular controls relevant to the design. Other documentation exists which will give guidance on what procedures should be included and at what frequency. Examples may include:

- *material standards tests, for example aggregate grading*
- *signed-off certification at hold points and of installation processes, for example rebar fixing*
- *concrete slump*
- *cored sections for lining thickness and concrete strength*
- *encountered ground conditions.*

9.5 Monitoring

The design of the lining will result in various inherent forms of behaviour during and post-construction. For example, it will deform as it comes under load and surface settlement will occur. The causes and effects of these are dealt with in detail in the relevant design sections. The importance of QC is to understand what should happen to the lining in accordance with the design and to monitor this behaviour and ascertain that the lining is performing as designed.

9.5.1 Lining deformation

All linings will have a degree of flexibility resulting in deformation of the initial built shape as it comes under load. This may take some time depending on the nature of the ground and any imposed loads, both internally and externally. The nature of the deformation will also vary depending on the lining type and ground conditions. Segmental linings will 'squat' in ordinary consolidated ground while in overconsolidated ground the lining may be squeezed reducing the diameter at springing level and increasing the vertical diameter.

The anticipated or acceptable level of deformation for a precast segmental lining will be dependent on the structural design of the ring and the connecting mechanism between segments. The quality of workmanship and method of construction may also influence it to a degree. Cast in situ and sprayed concrete linings should deform in a more predictable manner as a direct consequence of the structural design.

The prediction of such deformation is covered in Chapter 7 and Section 5.3, however, the anticipated or acceptable levels of deformation should be an output of the Design and Manufacture Quality Plans, and should become a requirement under the Construction and Post-construction (Operation) Quality Plans.

It follows that a system to monitor and record deformation against time is required to demonstrate compliance with the design requirements. In many cases this will be a simple series of diametric measurements at regular spacing and prescribed time intervals along the length of the constructed tunnel.

Spacing may, for example, be arbitrarily set as:

- *each ring immediately after erection*
- *each ring after 'exiting' the tail-skin of the shield*
- *every fifth ring after full ground loading has occurred*
- *every 100 m after completion and all live loading has occurred.*

Alternatively, spacing may be designed to match changes in ground conditions where predicted levels of deformation may vary.

Time-scales will be dependent on the rate of advance, the nature of the ground to fully load the tunnel, and the application of live loads.

Measurements will include, as a minimum, the horizontal and vertical diameters. Additional measurements may be included where asymmetrical loading is creating similar deformation.

The Construction Quality Plan prescribes what is to be monitored, when and how it will be measured, and how the data will be recorded. It should also refer to thresholds of acceptability and trigger levels at which specified actions should be taken.

In the case of sprayed linings, where the results of monitoring may be used to confirm and/or vary future design processes, the monitoring programme may be more complex and also include, inter alia, strain and stress measurements. This process is more complex and covered as part of Chapter 8 of this Guide. The importance of a prescribed monitoring programme within the Construction Quality Plan remains paramount.

In all cases, the reporting of results in a specified manner within the Construction Quality Plan allows the data to be easily and consistently analysed and actions to be taken against set trigger levels. It follows that the system must provide sufficient accuracy to recognise departures from accepted limits but be sufficiently workable to provide results such that actions can be taken in a timely manner. Various software applications lend themselves to computerised analysis of data and graphical representation of results.

9.5.2 Surface settlement

Surface settlement is primarily a consequence of the method of construction. Little, if any, surface settlement could be attributed to the actual design of the lining, for example as a result of deformation.

Notwithstanding this distinction, the effects of settlement on surface and subsurface structures are a key consideration in the overall design of a tunnel project.

The prediction of settlement is covered under Chapter 7 and the determined figures will be an output from the Design Quality Plan.

The Construction Quality Plan will consequently contain the following requirements:

- *maximum anticipated settlement at defined points, for example tunnel centreline, sensitive buildings, etc.*
- *the procedure for monitoring actual settlement*
- *the method of recording and reporting measured data*
- *trigger levels at which specified action shall be taken, for example begin compensation grouting.*

The actual monitoring procedure can vary significantly from simple centreline levels at regular spacing, to designed grid levels and specific critical level locations. The process can be manual, or automated to a highly sophisticated standard, which may also include computerised recording, analysis and representation of results. The design of this system will be dependent on the anticipated settlement and the sensitivity of structures to this settlement.

The timing of the monitoring process is dependent on the rate of advance of the tunnel and on the behaviour of the soil. The procedure should include the following.

- **Recorded levels for a period of time before the tunnel excavation passes a monitoring point,** for example three readings at weekly intervals to identify any natural ground movement.
- **Recorded levels at a specified distance in front of the tunnel face as it arrives at a monitoring point.** This will record the development of the settlement trough (or potentially heave) ahead of the tunnel face and provide early information on how the settlement is developing compared with the anticipated behaviour.
- **Recorded levels at regular time intervals as the tunnel face passes a monitoring point.** For example, for a rail crossing this may be hourly intervals until no discernible movement is recorded. In less sensitive areas this may be every 12 hours until 75% of the anticipated settlement has occurred and at 24-hour intervals until all discernible settlement has been recorded.

Recognition should be given to the rate of settlement and predictions may be possible, based on this information, to check if the theoretical maximum settlement could ultimately be exceeded. This will allow action to be taken at prescribed trigger levels before the anticipated settlement is exceeded. This is particularly relevant to compensation grouting covered in Section 7.3 of this Guide.

10 Case histories

10.1 Heathrow Express – design and performance of platform tunnels at Terminal 4

10.1.1 Project background

The Heathrow Express project was conceived in 1986 as part of a study to look at improving access links between Central London and Heathrow airport. The rail link runs on the existing British Rail main line from Paddington to north of the M4 from where, there is 8 km of new alignment to serve stations in the Central Terminal Area and at Terminal 4 (T4).

The layout of the T4 station complex is shown in Fig. 10.1 and comprises two platform tunnels with a central concourse at one end. These are connected by a series of cross-passages and intersected by the North and South Ventilation tunnels at each end of the station.

10.1.2 Geotechnical

The geological sequence at Terminal 4 was based on site specific investigation data and consisted of the following:

- *Made Ground typically 2 m thick*
- *Terrace Gravels between 2 and 4 m thick*
- *London Clay to 60 m depth*
- *Lambeth Group underlying the London Clay.*

10.1.3 Design

Prior to the collapse of the Central Terminal Area (CTA) station tunnels, the Contractor, Balfour Beatty (BBCEL), with Geoconsult as designer, was responsible for the primary support, while Mott MacDonald (MM) was responsible for the permanent support system. Following the collapse, MM was responsible for the primary support and permanent support systems. Each design section was independently checked by Faber Maunsell in a manner consistent with a Category 3 check.

10.1.3.1 Design approach The linings were designed to satisfy the requirements of ultimate limit state (ULS) design according to British Standard BS 8110 (British Standards Institution, 1997).

Numerical analyses by MM were carried out using the FLAC (Fast Lagrangian Analysis of Continua) program, which uses an explicit finite difference formulation for the solution of 2D and 3D problems.

10.1.3.2 Ground Model For the analysis, the Made Ground and Terrace Gravels were assumed to have the same properties. These were characterised by an elasto-plastic model with a Mohr–Coulomb failure criterion.

For the London Clay, a soil model was developed and implemented into the FLAC program. The soil model took into account the fact that the stiffness of London Clay is highly strain dependent and the degree of non-linearity depends on factors such as stress levels and the stress history of the soil. General parameters adopted in the model are shown in Fig. 10.2 and Figs 10.3a, 10.3b and 10.3c.

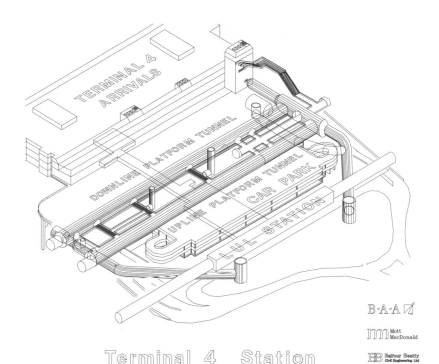

Fig. 10.1 Heathrow Express, Terminal 4 layout

Terminal 4 Station

B·A·A

mm Mott MacDonald

BB Balfour Beatty Civil Engineering Ltd

Parameter	Made ground and Terrace Gravels	London Clay
Bulk density: kN/m³	20	20
C': kN/m²	0	0
ϕ': degrees	30	25
C_u: kN/m²	–	see Fig. 10.3a
E: MN/m²	75	see Fig. 10.3c
Poisson's ratio	0.3	0.49
K_0	0.9	see Fig. 10.3b

Fig. 10.2 Soil parameters

The model assumes that both undrained strength and small strain stiffness are isotropic. Failure in compression is assumed at approximately 1% strain. Beyond the peak strengths in both compression and extension, the post peak stress–strain behaviour is adopted such that the shear stress reduces to an ultimate value equivalent to 0.5 times the peak shear strength.

For long-term effective stress analyses the London Clay soil model takes the initial small-strain shear modulus as proportional to the undrained shear strength with the undrained shear strength adopted calculated as 50% of the intact value to account for long-term creep effects. A constant value of 0.025 is taken for the slope of the voids ratio (e) versus effective stress ($\log p'$) plot to calculate bulk modulus.

10.1.3.3 Analysis A surface surcharge of $85\,\text{kN/m}^2$ was applied across the width of the model to simulate the additional load due to the surface structures within the T4 area.

The modelling sequence adopted for the assumed staged construction of the tunnels was selected to represent as faithfully as possible the actual construction stages.

The Hypothetical Modulus of Elasticity (HME) soft lining approach was adopted to take account of the three-dimensional face effect and the 'green' shotcrete behaviour, and to allow

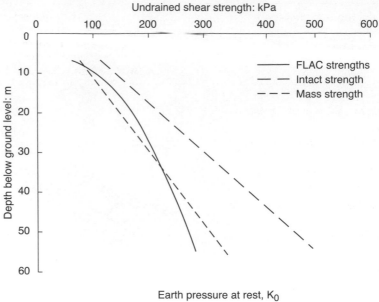

Fig. 10.3a Undrained shear strength

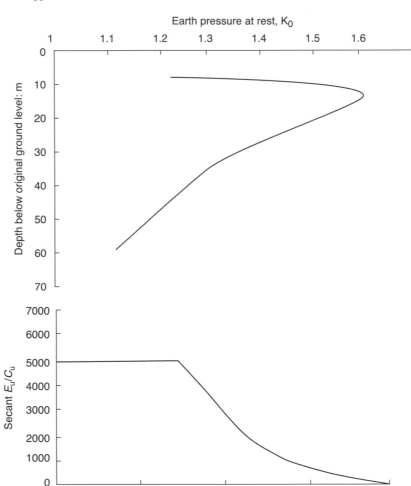

Fig. 10.3b Coefficient of earth pressure at rest

Fig. 10.3c Variation of compressive secant modulus of elasticity normalised by undrained shear stress (E_u/C_u) versus total strain

deformations to occur prior to installation of the stiffened shotcrete shell. The HME stiffnesses were selected to achieve, in tunnel deformations, an equivalent to a face loss similar to the maximum face loss observed at the Heathrow Express (HEX) Trial Tunnel. Surface settlement data from the HEX Trial Tunnel was used to validate the numerical model.

The modelling sequence adopted is shown in Fig. 10.4.

The primary shotcrete linings were modelled as connected rigid 'weightless' beam elements with a Young's Modulus equivalent to the HME.

Tunnel lining design guide. Thomas Telford, London, 2004

Tunnel modelling stage	Shotcrete stiffness (HME) in GPa					
	Platform tunnel 1			Platform tunnel 2		
	Top heading	Bench	Invert	Top heading	Bench	Invert
Stage 1 Excavate 1st tunnel top heading	0.75	–	–	–	–	–
Stage 2 Excavate 1st tunnel bench	2.0	0.75	–	–	–	–
Stage 3 Excavate 1st tunnel invert	2.0	2.0	0.75	–	–	–
Stage 4 Excavate 2nd tunnel top heading	25	25	25	0.75	–	–
Stage 5 Excavate 2nd tunnel bench	25	25	25	2	0.75	–
Stage 6 Excavate 2nd tunnel invert	25	25	25	2	2	0.75
Long-term analysis	25	25	25	25	25	25

Fig. 10.4 Modelling sequence

The excavation stages were carried out using undrained soil parameters. The model assesses porewater pressure changes due to the imposed loading conditions and deformations. Long-term effective stress analyses were completed assuming that the lining was fully drained or acted as an impermeable membrane giving rise to full hydrostatic pressures.

10.1.4 Lining details

The platform tunnel had external dimensions of 8.33 metres in height and 9.24 metres in width, the lining geometry and shotcrete thickness are shown in Fig. 10.5.

The ability of the shotcrete lining to carry the loads was checked at the Ultimate Limit State (ULS) in accordance with BS 8110 (British Standards Institution, 1997). Envelopes of the allowable load bearing capacity for the appropriate thicknesses and shotcrete strengths were checked against the FLAC output.

10.1.5 Instrumentation and monitoring

Instrumentation sections were established at intervals along the tunnel centreline. Three different types of monitoring sections were developed as follows.

- **Regular Monitoring Section (RMS)** – spaced at 10 m intervals and comprised six convergence bolts with optical targets.
- **Stress Monitoring Section (SMS)** – spaced at 20 m intervals and comprised radial and tangential pressure cells and piezometers, in addition to the convergence bolts.
- **Main Monitoring Section (MMS)** – spaced at 40 m intervals. In addition to the instrumentation of an SMS, the MMS included extensometers and inclinometers installed from the surface.

The purpose of this monitoring was first to verify that the tunnels were behaving in line with the design assumptions and second to detect any anomalous behaviour early enough so that counter-measures could be taken. The monitoring results were reviewed on a daily basis and compared to the control limits (e.g. for lining convergence), which had been determined from the design analyses.

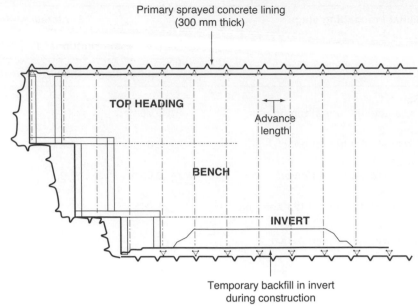

Primary sprayed concrete lining
(300 mm thick)

TOP HEADING

Advance
length

BENCH

INVERT

Temporary backfill in invert
during construction

LONGITUDINAL SECTION

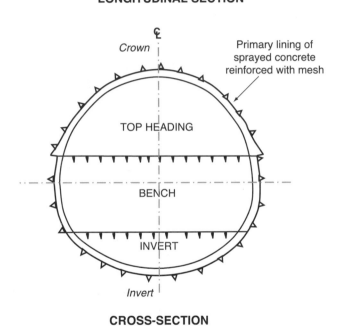

℄

Crown

Primary lining of
sprayed concrete
reinforced with mesh

TOP HEADING

BENCH

INVERT

Invert

CROSS-SECTION

Fig. 10.5 Platform tunnel
geometry

The whole process worked well in practice. The overall volume
loss for the platform tunnels was about 0.90%.

10.2 Design of Channel Tunnel lining

10.2.1 Project history

The concept of a tunnel linking Great Britain and France goes back to Napoleonic times. The first attempt at construction of a channel tunnel was in the 1880s by Sir Edward Watkins of the London and South Eastern Railway Company. The British military authorities, who feared invasion, stopped his attempt. A further attempt was made in the 1920s when a trial bore using the Whitaker Machine was started at the Folkestone Warren.

In 1955, the British Minister of Defence declared a channel tunnel was no longer a threat to national security and in 1974 construction of a bored rail tunnel began on both sides of the channel. Again the British Government stopped this as it lacked the commitment to the necessary financial guarantees.

In 1980, the British Government invited the private sector to provide ideas and proposals for the construction of a fixed link. Formal invitations to bid were issued in 1985 and by early 1986 the Channel Tunnel scheme of a rail shuttle service for road vehicles with provision for through trains, was accepted.

10.2.2 Design background

When the project first started in the 1970s the tunnelling works from the late 1800s were able to be referenced and examined. Both British and French engineers at the time thought that a visco-elastic model for the chalk marl could be used to design the bored tunnels.

Even though the project was cancelled in early 1975 a 250 m length of TBM service tunnel, which included instrumented lengths of tunnel and its segmental linings, was allowed to proceed. The location of these lengths could be reached by short headings from the 1880 tunnel in advance of the arrival of the TBM. The early results were reported in 1976. Mott, Hay and Anderson monitored the instrumentation subsequently, mostly at their own expense.

By the revival of the project in 1985, it was clear that the 1975 instrumentation was no longer very reliable. A significant number of the gauges were not performing adequately, but they could do so and were used successfully later to examine the short-term effects of the passage of the running tunnels on the existing service tunnel as those tunnels passed their location.

The experiences of the tunnelling in the 1970s project were different in the drives from France and the UK. In France the tunnel was accessed from an inclined adit along the line of the tunnel, which proved very difficult to construct owing to the heavy inflow of groundwater. It should also to be realised that the strata at tunnel level on the UK side of the Channel varied only gradually, but that near to the French coast the strata became much more convoluted, such that it was not possible to keep the tunnel within the preferred Chalk Marl horizon. By the time of the 1985 project the French constructors wished to take advantage of recent improved developments of pressurised shield TBMs with bolted and gasketted segmental linings. These would slow-down progress, but be less affected by groundwater inflows than the UK decision to use open shield TBMs with expanded segmental linings, which could be built very quickly. In the event the UK constructor did eventually achieve record-breaking advance rates, but shortly after the start of the UK tunnel drives water inflows and poor ground delayed progress for about a year. The final outcome was that the average progress rates were very similar.

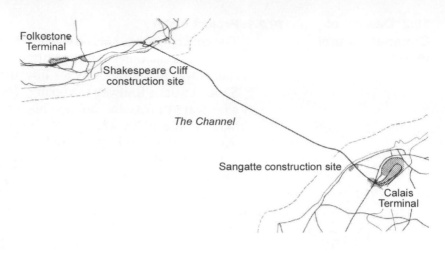

Fig. 10.6 Channel Tunnel
route

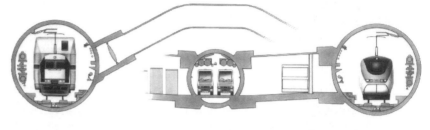

Fig. 10.7 Cross-section
through the three-tunnel
system

The principle adopted was that the constructors from each side of the Channel should be allowed to build in those ways in which they were most confident. However, both the French and UK designers did use the same agreed Design Manual.

The route of the Channel Tunnel and its cross-section are shown in Figs 10.6 and 10.7.

10.2.3 Geotechnical

The Channel Tunnel lies within the northern limb of a large regional anticlinal feature known as the Wealden–Boulonnais dome as seen in Fig. 10.8. On the English side, the strata are relatively flat, lying with a dip of usually less than 5°, but towards the French coast this increases locally to 20°.

The lowest member of the Chalk Formation is a calcareous jointed mudstone known as Chalk Marl. The combination of chalk and clay within the horizon results in a relatively strong and stable material, generally free from open discontinuities and hence virtually impermeable. Above the Chalk Marl the Lower and Middle Chalks were known to be more brittle and fissured and the Glauconitic Marl and Gault Clay below, although impermeable, are more likely to undergo time-dependent deformation as a result of stress changes due to tunnelling.

The vertical profile seen in Fig. 10.8 was defined by the decision to keep the tunnel as far as possible within the Chalk Marl. The presence of a continuous good tunnelling stratum of Chalk Marl linking the UK with France made the Channel Tunnel technically feasible.

Several sections of the 1975 service tunnel had been monitored since their installation and back analysis of measurements provided a check on the values of key geotechnical parameters used in the design. The effects of ground strata interfaces and adjacent tunnel construction were determined using finite element techniques.

10.2.4 Summary of parameters

A summary of parameters is shown in Fig. 10.9.

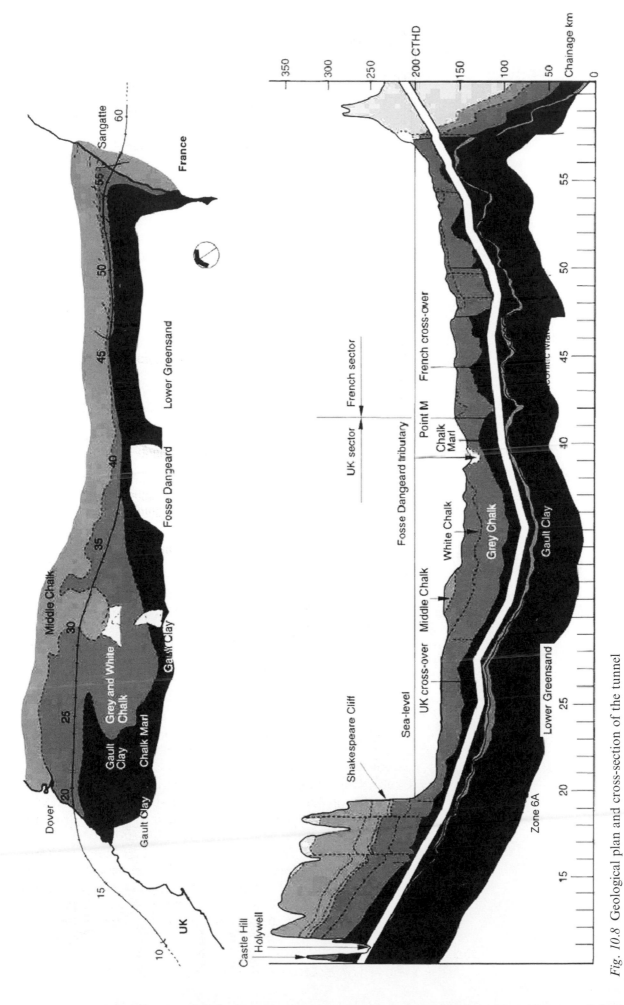

Fig. 10.8 Geological plan and cross-section of the tunnel

Parameters	Units	White chalk	Grey chalk	Upper Chalk Marl	Lower Chalk Marl	Glauconitic Marl	Tourtia	Gault Clay
				Stratum				
γ_{sat}	kN/m³	21.5	22.5	22.5	23.0	23.5	22.0	21.5
UCS	MPa		9.0	9.0	6.0	4.7	4.0	3.0
E	MPa		1300	1300	900	900	500	250
E_{50}	MPa		975	975	675	600	375	175
ϕ (creep ratio)			1.5	1.5	1.5	1.5	2.1	2.1
ν (drained)			0.3	0.3	0.3	0.3	0.3	0.3
ν (undrained)			0.5	0.5	0.5	0.5	0.5	0.5
Pre-peak								
c'	MPa		1.2	1.2	0.9	0.5	0.5	0.25
ϕ'	deg		35	35	35	30	30	30
Post-peak								
c'	MPa		0	0	0	0	0	0
ϕ'	deg		35	35	35	30	30	30
K_0 (longitudinal)			1.5	1.5	1.5	1.5	1.3	1.3
K_0 (transverse)			0.5	0.5	1.5	0.5	0.7	0.7
Mass permeability	m/s		1×10^{-9}	2×10^{-7}	7×10^{-8}	7×10^{-8}	1×10^{-9}	

Fig. 10.9 Parameters adopted in design of first part of UK undersea drive

Tunnel lining design guide. Thomas Telford, London, 2004

10.2.5 Lining design

The experience gained in the aborted project of the 1970s led to the conclusion that the excavation of the main tunnel drives should, for economy and speed, be carried out by fully mechanised machines, and that the tunnels should be lined with pre-formed structural elements erected immediately behind the machines.

10.2.5.1 Design criteria Under the contract, the design was required to satisfy the following provisions.

- **Design life** All of the permanent works are to be designed for a life of 120 years.
- **Fixings** Items such as fixings and internal caulking to have a design life of 25 years and be accessible.
- **Loadings** The calculations shall conform to two separate limit states; the first being the standard design cases taking account of the interaction between the lining and the ground, the in situ state of stress in the ground, the deflection of the linings and the redistribution of the loading dependent upon the relative flexibility and compressibility of the lining; the second being to ensure that the ultimate limit state of the lining is checked against the unfactored full ground and water overburden. The linings were checked for seismic loads.
- **Watertightness** There shall be no dripper (1 drip/minute) in the upper half of the tunnel onto sensitive equipment. There shall be no continuous leaks (>4 litre/h) other than those being diverted directly into the drainage system.
- **Circularity** The maximum value of lack of circularity, due to movement underground both as a result of loads and building tolerance, shall not exceed 1% of the radius.

10.2.5.2 Ground loading The design method adopted for the segmental tunnel linings was derived from studies carried out for earlier projects. It was determined from the studies that the visco-elastic model could be used to design the segmental tunnel linings. This was essentially an empirical method calibrated from the evidence of the previous century of data acquisition, and was refined over the course of the project.

The simplest model employed the Kelvin visco-elastic theory. For uniform radial loads this can be expressed as:

$$N = [P_0]\left[(1 - \lambda) + \lambda\left(e^{-\gamma T_0}\frac{\phi}{1 + \phi}\right)\right]\left[\frac{1}{1 + Q}\right][R] \qquad (10.1)$$

where N is the hoop load acting on the lining, λ is the deconfinement ratio, γ represents a rate of creep of the ground, ϕ is the ratio of creep to immediate (elastic) strains in the ground, T_0 is the time delay between excavation and lining, P_0 is the pre-existing radial stress in the ground, Q is the ratio between the elastic constants of the ground and lining (E, ν), modified for creep effects, multiplied by the radius R divided by the lining thickness.

This model can be extended to anisotropic stress fields and to plastic ground behaviour; this has been done in the detailed design stage. In the above formula, values for λ were taken as 1.0 for the UK TBM drives and 0.6 for cross-passages.

The value of the term $e^{-\gamma T_0}$ was usually assumed to be 1.0.

For watertight linings the calculation of ground load is made using effective stresses and water loading is computed separately from the relationship using equation (10.2).

$$N = [P_w]\left[\frac{1}{1+Q}\right][R] \tag{10.2}$$

The precast linings in the UK drives are not watertight and water pressures are dissipated and transferred to the surrounding ground. As a first approximation, the ground and water loads can be combined and their effects calculated.

Finite-element analyses were carried out in parametric studies to establish the importance of various parameters upon the designs. In general, two-dimensional analyses incorporated visco-elastic and/ or visco-plastic models, whereas the three-dimensional analyses required for the design of openings and junctions usually used a linear elastic model.

Studies were carried out to determine *inter alia* the effects of:

- *ground strata interfaces*
- *subsequent tunnels upon the first*
- *a running tunnel on a previously constructed cross-passage, etc.*
- *junctions upon the main tunnel linings.*

Service and running tunnel sections are shown in Fig. 10.10.

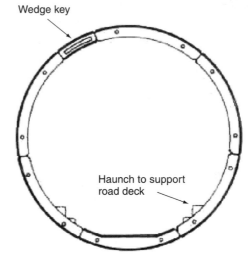

Service tunnel precast concrete lining
Nominal 4.8 m internal diameter

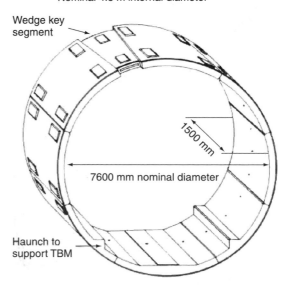

Fig. 10.10 Service and running tunnel sections

Typical running tunnel lining

Tunnel lining design guide. Thomas Telford, London, 2004

10.2.6 Precast segmental lining design

For reasons of economy, rings of expanded and articulated precast concrete segments were the preferred choice of lining for the majority of the main UK tunnel drives, 442,755 such segments being produced. In the French drives, articulated, bolted and gasketted linings were adopted to suit the predominantly wet and fissured ground conditions.

In developing the design of the precast concrete (PCC) linings, the scale of the project, the need for rapid progress both in establishing the supply of segments and during construction in the tunnel, and an obligation to use the best current practice in workmanship and materials were all taken into consideration. Figure 10.11 shows the details of the 1.5 m long and 270–540 mm thick lining segments.

These figures indicate some of the ways in which the casting facility was used to provide features to simplify handling and to speed construction. For example, the integral grout-pads to aid ring building, and to ensure a continuous path for cavity grouting, posed a technical challenge for the supplier. The holes for fixing permanent equipment are another example of the intentions to pursue the one-pass lining approach as far as possible. Other details not shown in the figure include rebates at the ends of mating circumferential and radial joint surfaces to minimise the risk of damage during construction in the tunnel.

It was necessary to provide steel reinforcement in such large segments primarily for safety during handling and erection, and to resist tensile bursting stresses near the articulated radial joints, which cause concentrations of the compressive hoop load in the

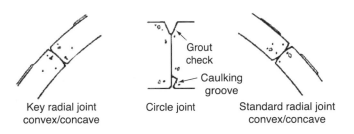

Key radial joint convex/concave   Circle joint   Standard radial joint convex/concave

Typical joint details

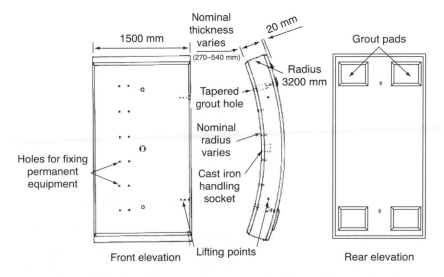

Fig. 10.11 Precast segment details

Typical arrangement of precast concrete segment

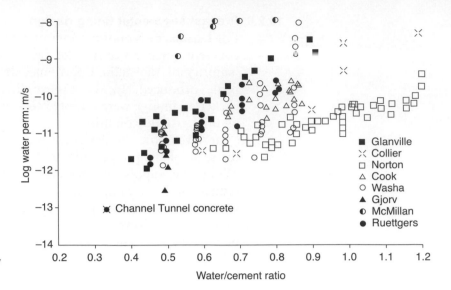

Fig. 10.12 Comparison of concrete mixes showing the Channel Tunnel concrete has exceptionally low permeability and diffusibility

rings. This in turn raised the problem of durability due to corrosion of the reinforcement, which was a serious concern in view of the saline nature of the groundwater expected in the undersea tunnels. In order to control the corrosion problem the following measures were taken:

- **Concrete mix**   A concrete mix of exceptionally low permeability and diffusibility was provided as illustrated in Fig. 10.12.
- **Reinforcement cover**   Steel reinforcement was covered to 35 mm, a compromise between the conflicting demands of durability and strength of the segments, which is compromised if the cover/ thickness ratio is too large.
- **Annulus grouting**   The annulus between the lining and the excavated ground was cavity grouted, together with proof-grouting as necessary to minimise water inflows.
- **Drainage paths**   Closed drainage paths were constructed for such water that might penetrate the joints between segments.
- **Reinforcement cages**   Reinforcing cages were fabricated by welding, and were provided with bonding terminals for possible future localised cathodic protection schemes.

Permeability and diffusivity tests were carried out during the mix development, and during production. In view of the quality of the concrete, the tests took a long time to carry out. Drilling holes and testing the powder produced detected the advancing chloride front. A computer program was developed and employed to estimate the time at which critical concentrations of chloride would reach the reinforcing bars.

Reinforcement at joints is provided to suit the design load as derived from small- and large-scale laboratory tests, employing the design chart shown in Fig. 10.13. The load-carrying capacity of precast concrete linings was adjusted by the amount of steel reinforcement included. At junctions with cross-passages and other areas of locally enhanced loads, ductile SGI was employed in order to avoid a need to increase the tunnel size everywhere to accommodate such loads.

10.2.7 SGI lining design
In the UK drives, it was also necessary to provide alternative bolted SGI linings for ground conditions in which an articulated expanded concrete lining could not be built to satisfactory standards.

Tunnel lining design guide. Thomas Telford, London, 2004

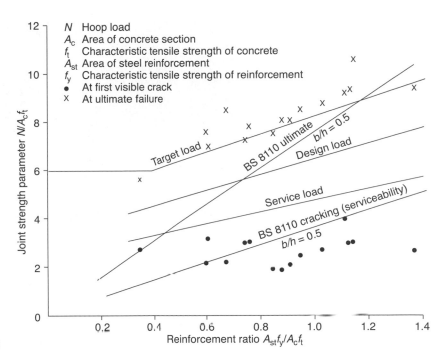

Fig. 10.13 Design chart compiled from joint test results

Although there have been a few previous applications in tunnels of SGI those had mainly been based upon the substitution of SGI for the brittle Grey Iron used over the previous hundred years. The tensile strength and ductility advantages of SGI had rarely been used. A research project undertaken in the mid-1980s showed the potential for cost savings, and further tests commissioned by the constructor, TML, served to establish sensible design parameters. These were adopted in the project but, except for hand-mined tunnels such as cross-passages and pressure relief ducts between the main TBM drives, economics dictated that the use of SGI should be minimised in the main drives. Spheroidal graphite iron is however a superb construction material, particularly for tunnel segments. Comparisons between SGI and high tensile steel are shown in Fig. 10.14.

In order to guard against corrosion a 1 mm allowance was added to the thicknesses of the skin and peripheral circular flanges of cast-iron segments. This does not necessarily provide for 120 years of deterioration, but it gives time to discover deterioration and to

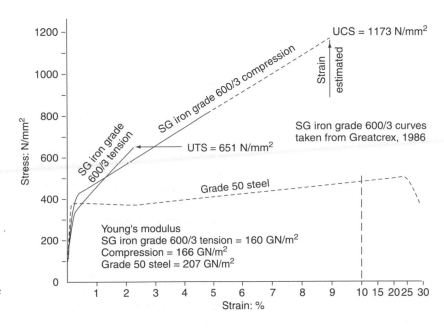

Fig. 10.14 Stress versus strain for grade 600/3 SGI and grade 50 structural steel

Fig. 10.15 Technical room lined with SB3 SGI rings

take measures to counteract it. All surfaces were coated with a bituminous paint.

The appearance of the bolted cast-iron linings is shown in Fig. 10.15. The rings designed for the main tunnels were built using segments, which could be manhandled if necessary. For mechanical erection, panels of four segments were bolted together on the surface and then transported and erected as one piece. All flanges were machined and each incorporated a groove to accommodate a rubber gasket (EPDM) if required. Linings designed for use in hand-mined tunnels were generally provided with a smaller groove into which a hydrophilic gasket could be inserted.

Initially it was anticipated that, at junctions between the main tunnels and cross-passages and ducts, several rings of SGI linings would be placed in the main tunnels. The original concept was that rings forming openings in the main tunnels should be built one-by-one without interruption. The constructor, TML, developed this concept further in the UK, whereby the construction and erection of all segments should not differ, whether or not their principal material was concrete or cast iron. From this the 'hybrid ring' was developed. In the hybrid openings, the use of SGI segments was limited to segments immediately adjacent to the openings.

Pressure relief ducts between the main train-running tunnels have to pass over the intermediate service tunnel at regular (250 m) intervals. For aerodynamic reasons these tunnels need a smooth internal bore and, from construction considerations, a lightweight SGI lining was desirable. Thus, TML proposed a lining in which the circumferential skin was entirely in the intrados and required an elaborate system of temporary connections. Despite the fact they were built in difficult circumstances, construction of the ducts proceeded with great success.

10.3 Great Belt railway tunnels

10.3.1 Plan, geotechnical longitudinal section and cross-section

The Storebælt railway tunnel provides the fixed link across the Eastern Channel between Zealand and the small island of Sprogø in Denmark. The 7.7 m diameter running tunnels are separated at 25 m centres. The maximum tunnel depth is 80 m below mean sea level.

Figures 10.16 and 10.17 show the tunnel's plan and longitudinal profile and its lining cross-section.

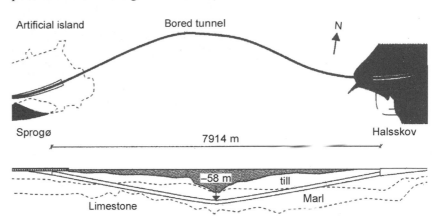

Fig. 10.16 Plan and longitudinal profile

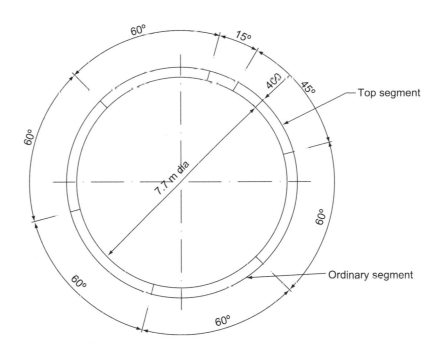

Fig. 10.17 Lining cross-section

10.3.2 Geology

The Eastern Channel is an ancient glacially eroded channel, increasing steadily in depth to 20 m before rapidly steepening to a central channel, up to 55 m deep. Site investigations showed complex ground conditions, resulting from the geological history of the area, and three bands of rock through which the tunnel would pass. The Quartenary glacial tills comprise clay with layers of silt and sand, meltwater deposits and boulders of granite and gneiss up to 3 m in diameter. Two bands were identified within the strata: an upper and lower till.

- **The upper till** – is the more uniform of the two and contains isolated sand deposits totalling less than 1% of the total mass. Undrained shear strength is 100–700 kPa.

- **The lower till** – is less homogeneous having been shaped by earlier glaciations. Sand and gravel deposits total up to 20% of the total mass. Undrained shear strength is 200–700 kPa.

The underlying upper Palaeocene Marl is a weak to moderately strong rock with fissured and jointed zones. The central channel has very fissured and jointed zones.

10.3.2.1 Outline of the geotechnical and geophysical investigations
The Danish Geotechnical Institute carried out site investigations for the Danish Government in five campaigns beginning in 1962 and ending in 1988. A total of 58 exploratory boreholes and vibrocores were performed and supplemented by extensive seismic surveys along a 200 m wide tunnel alignment corridor across the Eastern Channel of the Storebaelt.

During construction the client undertook further supplementary investigations in areas where information was inadequate or uncertain. Exploratory borings were also carried out when necessary, ahead of the tunnel boring machines (TBMs) and prior to the construction of each cross-passage.

The pre-contract investigations were not only reported in an Interpretative Geotechnical Report, but definitive properties of the strata materials were included in the Basis of Design Document (BDD), which was a contractual document.

10.3.3 Summary of geotechnical and geophysical properties
A summary of the geotechnical and geophysical properties is shown in Fig. 10.18.

Property	Symbol	Typical range	Average	Design value
Natural moisture content	W_{nat}	10–13%	11.5%	11.5%
Liquid limit	W_L		16%	16%
Plasticity index	I_p	5–7%		6%
Consistency index	I_c		0.70	0.70
Clay content	L		15%	15%
Activity index	I_A	0.35–0.50		0.40
Grain unit weight	γ_s		26.8 kN/m^3	26.8 kN/m^3
Bulk density	γ		22.9 kN/m^3	22.9 kN/m^3
Void ratio	e		0.31	0.31
Mean grain size	δ_{50}	0.05–0.10 mm	0.073 mm	0.07 mm
Undrained shear strength from vane tests	C_u	100–700 kPa		200–800 kPa
Effective strength parameters	c'	10–50 kPa		(10 kPa, 34°) or
(c' and ϕ')	ϕ'	30–36°		(30 kPa, 32°)
Slake durability index	I_{d2}	50–75%	63%	60%
Modulus of elasticity	E	20–30 MPa		23 MPa
Pressuremeter modulus	P_r	3.5–10 MPa	6.3 MPa	
Modulus of consolidation	K_t	(15–40) MPa $+(1500$–$3000)\sigma'_{red}$		20 MPa $+2500\,\sigma'_{red}$
Swell	ε_{sw}	0–0.4%	0.2%	0.2%
Poisson's ratio	ν	0.25–0.35	0.3	0.3
Earth pressure at rest	K_0	0.40–0.50		0.46
Coefficient of permeability	K	10^{-7}–10^{-5} m/s		10^{-5} m/s

Fig. 10.18 Geotechnical parameters of Till 1

10.3.4 Design of tunnel linings

The determination to use an Earth Pressure Balance TBM led to the design requiring a bolted, segmental precast concrete one-pass lining, erected in a tail-shield, and the annular void was to be grouted as soon as possible. Feasibility studies suggested that a high-strength and durable watertight (gasketted) segmental lining, 400 mm thick, should suffice.

The design basis provided information on the geometrical requirements for the tunnel, the permanent, variable and accidental loading, and the materials. The design was to be carried out on the basis of acknowledged conservative calculations confirmed by experience and according to the latest international or national codes for concrete and steel structures where possible.

A view of the intrados of a tunnel segment is shown in Fig. 10.19.

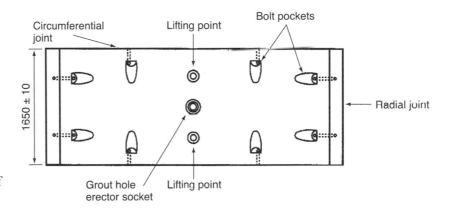

Fig. 10.19 View on intrados of segment

10.3.4.1 Analysis Finite element ground-lining interaction analyses were carried out in two dimensions. It was assumed that all elastic deformation had occurred before the lining was capable of restraining ground movements, and that subsequent deformation took place by visco-plastic ground behaviour. Results were obtained using both estimates and worst credible values for the various parameters.

The tills were modelled using standard Mohr–Coulomb strength criteria stiffness parameters, empirically derived assuming effective stress conditions. For the marl, the Hoek and Brown model was adopted, being more suitable for a brittle material in which strength falls suddenly from a peak to a lower residual value. A number of different situations were modelled (wholly in the tills, at the till/marl interface, and so on).

The results were treated with caution as very little plasticity arose and the modelling could not simulate discontinuities that might arise from sand lenses and other causes.

In effect the hoop load to be carried by the lining was readily determined, since the tills were known to be unstable, and the water loading predominated, and had to be carried by the lining. It was determined that a deformation of 1.0% could be achieved and should not be exceeded. Later, in discussions with the successful contractor, the number of segments in a ring was reduced, and a reduced deformation limit of 0.85% was agreed. A minimum compressive (cylinder) strength for the concrete of 60 MPa was agreed together with a minimum tensile strength for the purposes of the design of the radial (for long-term service loads) and circumferential joints (for construction loads to resist TBM shove forces).

10.3.4.2 Reinforcement It was necessary to reinforce the segments to avoid excessive cracking (see Fig. 10.20):

- *during handling and stacking of segments*
- *at radial and circumferential joints due to tensile and shear stresses generated by local stress concentration*
- *at bolt pockets.*

The amount of reinforcement was kept to a minimum, and was comprised of four elements:

- **Hoop reinforcement**   0.2% hoop reinforcement in each face, with half of this amount placed longitudinally to permit fabrication as a machine-welded mesh fabric
- **Ladder mats**   structurally welded 'ladder' mats in several layers at each radial joint to resist bursting forces
- **Circumferential joints**   through-thickness bars provided in quantity adjacent to circumferential joints but much less frequently elsewhere, to resist circumferential joint bursting forces and to provide a rigid cage
- **Local reinforcements**   local reinforcement around bolt pockets.

Fig. 10.20 Reinforcement cage for lining segment

10.3.4.3 Joints The radial joints were designed convex/convex to minimise eccentricity of hoop thrust, see Fig. 10.21. The corners of the plane-faced circumferential joints were chamfered to prevent damage by the TBM thrust rams, and bituminous packers were fitted to minimise the effects of uneven loading.

Joints were reinforced according to empirical results obtained from studies for the Channel Tunnel (Eves and Curtis, 1991).

10.3.4.4 Durability To achieve 100-year durability, a sulphate-resistant concrete for the segmental lining with a relatively low C_3A (tricalcium aluminate) content, ordinary Portland cement, microsilica and flyash additives was used. A very low diffusion coefficient of $600 \times 10^{-15}\,\mathrm{m^2/s}$ maximum, and permeability of $25 \times 10^{-15}\,\mathrm{m/s}$ maximum was aimed at to ensure a slow water transport, reduce expansion from sulphate attack, and bind chlorides.

Fusion-bonded epoxy coating of the fully welded complete reinforcement cages by the fluidised-bed dipping technique protected the reinforcement against corrosion and spalling of the concrete.

Tunnel lining design guide. Thomas Telford, London, 2004

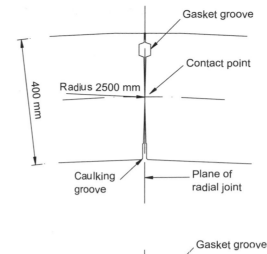

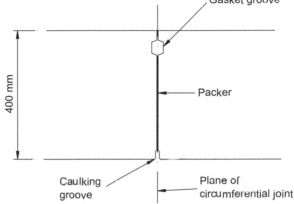

Fig. 10.21 Joint details

10.3.4.5 Watertightness The linings are designed to be watertight due to the occasional permeable ground in the tills and marlstones throughout the entire length in order to provide the required dry environment for the railway operating equipment and to ensure a 100-year durability.

A thixotropic grout was used in the annulus between linings and ground, with a relatively high C_3A content, to provide a measure of watertightness and bind some chlorides and sulphates.

The lining in all radial and circumferential joints close to the extrados of the lining was sealed with EPDM gaskets, which were designed to resist water pressures up to 16 bar maximum within joint movements and construction tolerances.

10.4 Instrumentation of the CTRL North Downs Tunnel

The following case history is based upon the 2002 Harding prize presentation by Hurt (2001) and the paper by Watson *et al.* (1999).

The North Downs Tunnel is a 3.2 km long, single-bore, twin-track rail tunnel having a 165 m^2 cross-section when excavated. It passes under the North Downs of Kent at up to 100 m depth. The tunnel was driven using SCL techniques through lower members of the White Chalk and the Lower Chalk, with the rock having intact uniaxial compression strengths in the range 2.0–4.5 MPa.

The tender design was developed by Rail Link Engineering (RLE), a consortium of design consultancies. The design-and-build contract was awarded based upon the ICE New Engineering Contract (Option C) (Institution of Civil Engineers, 1995). Subsequent design changes proposed by the contractor to offer programme and cost savings were subject to a rigorous checking and approvals process by RLE, and

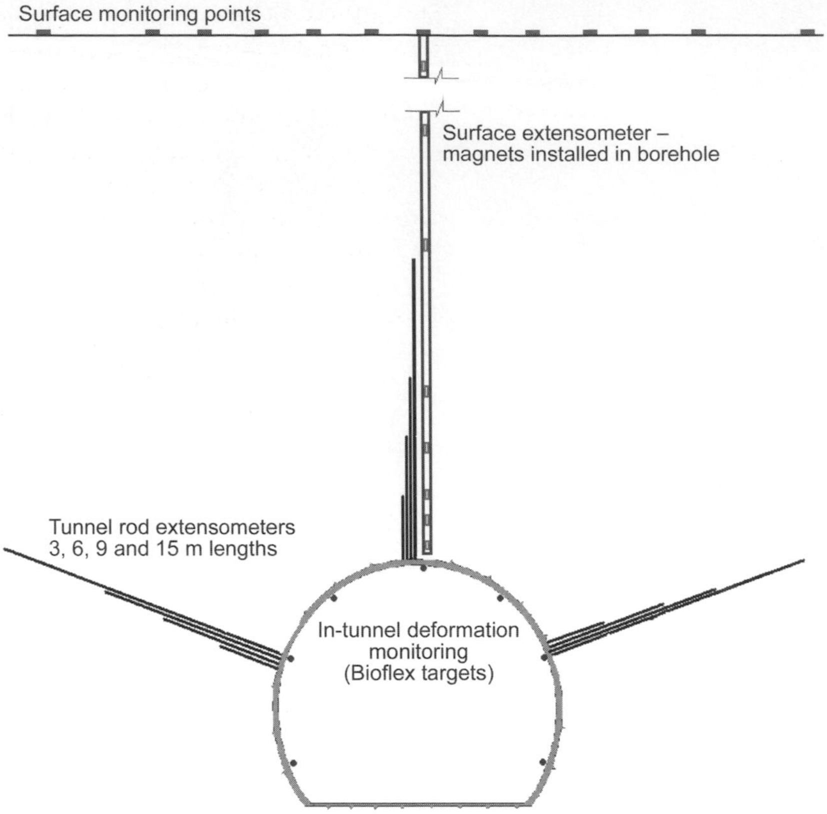

Fig. 10.22 Instrumentation used on the North Downs Tunnel

were then validated against the results of the monitoring during construction. A project management team from RLE supervised the construction of the tunnel.

A monitoring regime was adopted that provided information on the movements of the tunnel lining and on the ground around and above the tunnel. A cross-section of the tunnel illustrating the different monitoring systems is shown in Fig. 10.22.

Surface settlement was monitored using a total of 179 settlement points measured using precise levelling techniques. Settlement was measured on eight transverse arrays, mainly situated in the shallow cover areas, above the tunnel centreline. The transverse settlement measured during construction gave good approximation to a settlement trough following a Gaussian distribution where the cover to the tunnel was less than 30 m.

Surface extensometers were installed above the tunnel centreline in three deep boreholes and four further holes in shallow cover sections near the portals. There were five arrays of rod extensometers installed from within the tunnel. Each array consisted of three arms, one extending vertically up from the crown and the others at the base of the top heading, radial to the tunnel. The time taken to install the extensometers meant that the majority of the movement in the top heading was missed, but the extensometers in the deep sections gave good agreement with the tunnel deformation monitoring for bench and invert construction.

Monitoring of the in-tunnel deformation was carried out using precise three-dimensional surveys of arrays of Bioflex targets spaced along the tunnel at intervals of between 5 m and 40 m. Accuracy of the readings, which was governed by the quality of the targets and the atmospheric conditions in the tunnel, was ± 0.5 mm. The results of the deformation monitoring were post-processed using the Dedalos tunnel deformation programme.

Tunnel lining design guide. Thomas Telford, London, 2004

This allows the presentation of the vertical, transverse and longitudinal tunnel lining movements at each array, plotted against time.

The results of the monitoring were discussed at a daily support meeting, allowing adjustments to be made to the advance length and face support details. The results were also used to confirm the parameters adopted for the design of the Secondary Lining, with a back-analysis carried out at the location of each surface extensometer using numerical techniques. The back-analysis enabled validation of the in situ stress regime, the geotechnical parameters adopted for design, and the amount of relaxation assumed ahead of the tunnel face.

10.5 References

British Standards Institution (1997). *Structural Use of Concrete Part 1: Code of Practice for Design and Construction*. London. BS 8110.

Curtis, D. J. (1999). *The Channel Tunnel: Design Fabrication and Erection of Precast Concrete Linings*, Proc. Institution of Civil Engineers,

Curtis, D. J, and Bennick, K. (1991). Lining design for the Great Belt eastern railway tunnel. *Proc. Tunnelling '91 Conf.*, CIRIA, London.

Curtis, D. J., Eves, R. C. W., Scott, G. and Chapman, P. J. (1991). *The Channel Tunnel: Design, Fabrication and Erection of Precast Concrete Linings*, Tunnelling '91, IMMG, London, pp. 161–172.

Elliott, I. H. and Curtis, D. J. (1996). Storebaelt eastern railway tunnel: design. *ICE Proceedings*. Thomas Telford, London, pp. 9–19.

Hester, J. C., Crighton, G. S. and Curtis, D. J. (1991). The Channel Tunnel UK and France: The UK TBM Drives. *Proc. Rapid Excavation and Tunneling Conference, Seattle*.

Hurt, J. (2001). Primary ways to save – Harding Prize Paper. *Tunnels & Tunnelling International*, December 2001 and January 2002. Polygon Media, Sevenoaks.

Institution of Civil Engineers (1995). *The Engineering and Construction Contract. Option C: Target Contract with Activity Schedule*. Thomas Telford, London.

Institution of Civil Engineers (1992). The Channel Tunnel – Part 1 Tunnels. *ICE Proceedings*. Thomas Telford, London.

Norie, E. H. and Curtis, D. J. (1990). Design of linings for UK tunnels and related underground structures. *Proc. Conf. on Strait Crossings, Norway*. Balkema, Rotterdam.

Powell, D. B., Sigl, O. and Beveridge, J. P. (1997). Heathrow Express – design and performance of platform tunnels at Terminal 4. *Proc. Tunnelling '97*. IMMG, London, pp. 565–593.

Watson, P. C., Warren, C. D., Eddie, C. and Jäger, J. (1999). CTRL North Downs Tunnel. *Proc. Tunnel Construction and Piling '99*, September, Brintex/Hemming Group, London.

Appendix 1   Abbreviations and symbols

ABI	Association of British Insurers
ADECO-RS	Analysis of Controlled Deformation in Rocks and Soils
AGS	Association of Geotechnical Specialists
BDD	Basis of Design Document
BE	Boundary element
BGS	British Geotechnical Society
BRE	Building Research Establishment (UK)
BTS	British Tunnelling Society
CCM	Convergence–Confinement Method
CDM	Construction (Design and Management)
CIRIA	Construction Industry Research and Information Association (UK)
C_3A	Tricalcium aluminate
CTA	Central Terminal Area
CTRL	Channel Tunnel Rail Link
DE	Discrete element
DEMEC	De-mountable mechanical
DETR	Department of the Environment, Transport and the Regions (UK)
EDM	Electronic Distance Measurement
EPB	Earth pressure boring
EPDM	Ethylene Polythene Diene Monomer
EPSRC	Engineering and Physical Sciences Research Council (UK)
FD	Finite difference
FE	Finite element
FHWA	Federal Highway Administration (US)
FLAC	Fast Lagrangian Analysis of Continua
FS(O)	Full scale (output)
GBR	Geotechnical Baseline Report
GEMINI	GEotechnical Monitoring INformation Interchange
GRC	Ground Reaction Curve
GSI	Geological Strength Index
HAZOPS	Hazardous Operations
HDPE	High density polyethylene
HEX	Heathrow Express
HME	Hypothetical Modulus of Elasticity
HSE	Health and Safety Executive (UK)
I and M	Instrumentation and Monitoring
ICE	Institution of Civil Engineers (UK)
ISRM	International Society for Rock Mechanics
ISSMGE	International Society for Soil Mechanics and Geotechnical Engineering
ITA	International Tunnelling Association
JLE	Jubilee Line Extension (London, UK)
LVDT	Linear Variable Differential Transformer
MMS	Main Monitoring System
NATM	New Austrian Tunnelling Method
NEC	New Engineering Contract (ICE, UK)

Tunnel lining design guide. Thomas Telford, London, 2004

OM	Observational Method
PC	Personal computer
PCC	Precast concrete
PO	Polyolefin
PVC	Polyvinylchloride
QC	Quality control
QM	Quality management
QP	Quality Plan
RAM	Risk Analysis and Management
RLE	Rail Link Engineering
RMR	Rock Mass Rating
RMS	Regular Monitoring Section
RQD	Rock Quality Designation
RSST	Rapid shotcrete supported tunnel(ling method)
SCL	Sprayed concrete lined/lining
SEM	Sequential excavation method
SGI	Spheroidal Graphite Iron
SLS	Serviceability Limit State
SMS	Stress Monitoring Section
SRF	Stress Reduction Factor
T4	Terminal 4
TBM	Tunnel boring machine
TRRL	Transport and Road Research Laboratory (UK) – now Transport Research Laboratory (TRL)
ULS	Ultimate Limit State (design)
USBM	United States Bureau of Mines
$\mu\varepsilon$	10^{-6} strain or 'micro-strain'
VW(SG)	Vibrating wire (strain gauge)

Appendix 2   Risk management

A2.1 Introduction

No construction project is risk free. Risk can be managed, minimised, shared, transferred or accepted. It cannot be ignored. (Latham, 1994)

In recent years the civil engineering industry has identified the need for a systematic approach to risk management. There is now a growing body of literature which deals with the subject (see Bibliography). Knowledgeable clients now expect appropriate advice on risk management. Formal procedures need to be in place to deliver this advice in an effective manner. Experience shows that the application of systematic risk management procedures are particularly important for design-and-build projects.

The risk register is a formal method of recording threats and opportunities. The risk associated with each threat or opportunity is assessed in a logical manner jointly by an experienced team from the engineer, the Client and other key stakeholders and organisations involved in the project.

For most projects, a relatively simple qualitative assessment system is appropriate. For large projects or projects that are deemed particularly risky then a quantitative approach may be more appropriate. For ease of communication and subsequent tracking, the major risks identified by the assessment team should be summarised on a risk matrix.

A2.2 Scope

All projects should use a risk register. The level of sophistication and detail required will vary from project to project, depending upon particular requirements. This advice note and risk register is for project risk management, that is actively managing risks that may develop during the life of a project from inception through to completion, including any aspects that could impact on the operation and maintenance.

In use, the proposed risk register is intended to be:

- *as simple as possible*
- *applicable for all projects*
- *flexible and adaptable regardless of project complexity.*

It is recognised that some disciplines have specialist procedures for risk assessment and management, for example railway safety, contaminated land, seismic hazard, etc. Even if specialist processes are implemented, a simple high-level summary of risk for a project is still beneficial, particularly for a client who may not have the specialist knowledge of detailed quantitative risk analysis.

If used properly the risk management process can also deliver the following additional benefits:

- *promote sharing of knowledge and expertise in a cost-effective manner*
- *focus the project team's efforts on the critical issues*
- *improve communication across the team (thereby avoiding a 'silo' mentality).*

A2.3 Risk register

A2.3.1 When to use the risk register

It is intended that the risk register is a living document that all parties involved in the project use and update throughout the life of the project, particularly if events during construction require design modifications and adjustments to the programme. For design-and-build projects, the tender period is the critical phase, and it should be used to guide the tender process through to completion.

As a live document the risk register must be revisited, especially if new information becomes available or if there is a change in scope.

An example of a working register in given below.

A2.3.2 What is it?

The risk register is a summary of:

- *what can go wrong?*
- *how likely it is?*
- *what measures are required to mitigate the chance of something going wrong?*
- *who is responsible for managing it?*
- *who is vulnerable if something does go wrong?*
- *when risk management actions need to be carried out?*

It is also a communication tool, since it must be transmitted to *all* key members of a project team, including client staff and all key sub-contractors and sub-consultants.

A2.3.3 Assessment process

Careful thought must be given to the threat/opportunity identification phase, including:

- *technical background of individuals*
- *number of individuals who carry out the process*
- *the available background information.*

For most projects, it is suggested that between four and six experienced staff are required with a varied mix of general engineering, commercial and specialist technical backgrounds. Less than three and the breadth of experience and perspective would probably be inadequate. More than ten is difficult to manage effectively. Relatively short intensive 'brainstorming' sessions are usually most cost effective (typically 0.5 to 2.0 hours). Key background information on client requirements, contract conditions, geology, environmental issues, site constraints, etc. should be available for briefing key staff prior to brainstorming.

The prompt sheet, see Fig. A2.1, using key words/phrases is helpful in facilitating the identification of threats and opportunities for the project. They should be used before the end of a brain-storming session with the project team.

A2.3.4 Key steps

Figure A2.2 shows a flowchart which summarises the risk management process.

The stages are:

(i) threat/opportunity identification
(ii) assessment of likelihood
(iii) consequence analysis
(iv) consider options for reducing risk

Contract documents and company relations
- client – past working relationship
- customer's expectations
- sub-contractor/suppliers – competence, past working relationship
- contract conditions
- payment terms
- scope of works
- risk allocation – (e.g. ground risk)
- past design and build working relationship
- responsibility/authority boundaries
- communication lines
- handling information
- sensitivity to change (cost and time, alternative approaches)

Staffing
- staffing requirements
- relevant experience
- expertise involved at appropriate stage
- limitation of knowledge/expertise
- use of in-house specialists

Third parties and sensitivity
- Third party involvement
- reliance on other parties
- adjacent structures and services
- public involvement/concern
- location
- environmental issues
- aesthetics of finished work
- noise
- vibration

Approvals
- access
- regulations – environment, safety
- planning consents, licences
- client approvals
- waste management/minimisation

Ground conditions
- assessment of desk study, site investigation, interpretative report (are they adequate?)
- geological environment – potential variability, potential hazards
- hydrogeology – seasonal changes, long-term changes
- groundwater control
- contamination
- soil/structure interaction issues
- ground/structure movement
- earthworks

Design
- clear, unambiguous design brief
- serviceability criteria
- innovations or proven technology/methods/materials
- design interfaces
- adequacy and reliability of incoming data
- unforeseen mechanisms
- robustness of solution – design, workmanship, assumptions

Construction
- past experience with proposed methodology
- on-site verification/problem identification
- buildability
- maintenance
- innovations or proven technology/methods/materials
- instrumentation/monitoring
- construction interfaces
- feedback to verify design assumptions
- potential for observational method
- influence of changes to ground conditions
- temporary works

Programme
- sequencing of works
- time available
- access constraints
- availability of staff/specialist plan

Fig. A2.1 Prompt list for threat and opportunity identification

Tunnel lining design guide. Thomas Telford, London, 2004

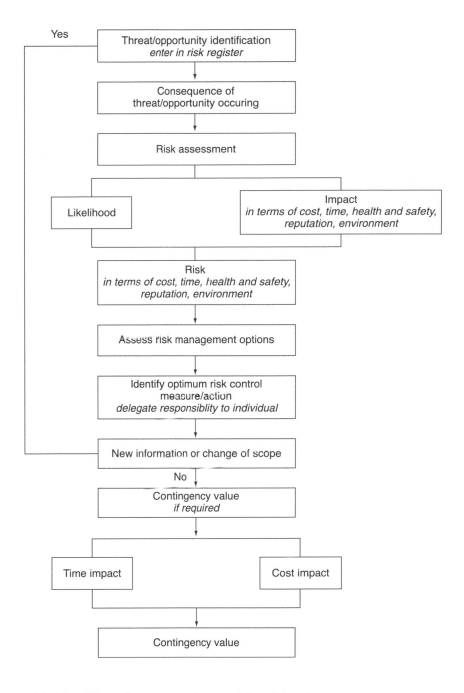

Fig. A2.2 Risk management process

(v) identify and manage chosen risk mitigation strategy

(vi) review and reconsider at regular intervals, especially if new information becomes available.

The most important step is threat/opportunity identification, if this is not carried out in a comprehensive manner then all subsequent stages are flawed.

For simple risk assessments stages (ii) and (iii) are qualitative and based on the combined experience and knowledge of the attendees; the main objective being to identify the most significant threats and opportunities for the project.

A2.3.5 Risk assessment, qualitative or quantitative?

The risk assessment can be either qualitative or quantitative depending on the project size, technical complexity, and preference of the parties involved. A qualitative risk assessment is likely to be appropriate for most projects. A full quantitative assessment can be advantageous in that it can be directly linked to cost and programme implications. However, unless risk specialists are available,

Impact		Cost	Time	Reputation	H and S	Environment
1 Very low	Negligible	Negligible	Negligible effect on programme	Negligible	Negligible	Negligible
2 Low	Significant	>1% Budget	>5% Effect on programme	Minor effect on local company image/business relationship mildly affected	Minor injury	Minor environmental incident
3 Medium	Serious	>5% Budget	>10% Effect on programme	Local media exposure/business relationship affected	Major injury	Environmental incident requiring management input
4 High	Threat to future work and client relations	>10% Budget	>25% Effect on programme	Nationwide media exposure/business relationship greatly affected	Fatality	Environmental incident leading to prosecution or protestor action
5 Very high	Threat to business survival and credibility	>50% Budget	>50% Effect on programme	Permanent nationwide effect on company image/significant impact on business relationship	Multiple fatalities	Major environmental incident with irreversible effects and threat to public health or protected natural resource

Fig. A2.3 Impact score

Tunnel lining design guide. Thomas Telford, London, 2004

undue effort on consequence and likelihood assessments can detract from the main purpose of the exercise that is to:

(i) identify the most significant risks
(ii) assess how best to manage these risks.

For current purposes a full quantitative approach is not recommended. Semi-quantitative and qualitative approaches are described below.

A2.3.5.1 Semi-quantitative approach Once the threats and opportunities have been identified and entered on the risk register then the associated risks are assessed. In order to derive a risk, the impact and likelihood for each threat or opportunity is considered. A few useful definitions are:

- **likelihood** – the chance (or probability) of the risk event occurring within a defined time period. Here the risk event is defined as either the threat occurring or the opportunity being lost
- **impact** – the effect of the risk event on one or more objectives if it occurs. The effect could be measured in accident rates, financial value, project delay in weeks, lost turnover due to damage to reputation, etc.
- **risk** – the potential occurrence of a threat or opportunity, which could affect (positively or negatively) the achievement of the project objectives
- **risk = impact × likelihood** – therefore, a high impact but very low likelihood would result in a low risk.

The scoring for the impact is given in Fig. A2.3 and is based on five categories: cost, time, reputation and business relations, health and safety, and environment. Impacts range from 1 (i.e. negligible impact) to 5 (i.e. catastrophic impact). It is intended that each threat and opportunity will be assessed for every category.

The risk score is derived by combining the impact and likelihood scores (from Figs A2.3 and A2.4) giving a risk level from trivial to intolerable (see Fig. A2.5).

		Likelihood	Probability
1	Very low	Negligible/improbable	<1%
2	Low	Unlikely/remote	>1%
3	Medium	Likely/possible	>10%
4	High	Probable	>50%
5	Very high	Very likely/almost certain	>90%

Fig. A2.4 Likelihood score   Note *Likelihood* means likelihood that a threat occurs/opportunity is lost.

A2.3.5.2 Qualitative approach The process for recording and assessing risks is similar to the semi-quantitative approach. However, the likelihood, impact and hence risk are scored as very low, low, medium, high or very high.

Figure A2.6 contains a project risk register for a qualitative approach. The column for risk type should contain a letter relating to the type of risk, for example health and safety or cost.

A2.3.6 Managing risk
Ultimately the success of a project depends on how well the risks are managed, in terms of:

		Impact				
Likelihood		Very low	Low	Medium	High	Very high
		1	2	3	4	5
Very low	1	n	n	n	n	t
Low	2	n	n	t	t	s
Medium	3	n	t	t	s	s
High	4	n	t	s	s	i
Very high	5	n	s	s	i	i

Key

intolerable (red)	i	15 to 25
significant (amber)	s	10 to 14
tolerable (yellow)	t	10 to 14
negligible (green)	n	

Fig. A2.5 Risk score matrix

Note: Insert ID number of selected threats/opportunities into risk matrix.

- *avoiding*
- *transferring*
- *reducing*
- *sharing the risks as appropriate.*

Following the risk assessment a joint decision should be made concerning optimum risk control measures and who 'owns' the risks. Note that the risk 'owner' is the organisation that is contractually liable if the risk occurs. The risk register allows the owner and risk control measure to be noted as well as the resulting residual risk. Many risks will be considered 'minor' and not require specific action. However, the assessment groups should decide on how many require specific actions and need to be closely monitored, perhaps the top six or ten risks together with the top two or three opportunities would be reasonable for many projects. Nevertheless, all 'RED' and 'AMBER' risks must be tracked. 'RED' risks, that is those risks which are considered 'intolerable' require a major policy decision, for example for design-and-build projects if an appropriate risk mitigation measure cannot be identified then a decision to 'not bid' may be required. The assessment team may also wish to track 'YELLOW' risks and take actions to move these towards 'GREEN'. The risk register contains a column for ranking the risks and marking those that are to be tracked.

The main purpose of individual risk scores or qualitative descriptions is to prioritise actions in a systematic manner.

Once the risks have been evaluated a decision must be made on how to:

- *remove the risk*
- *decrease the risk*
- *transfer the risk to another organisation*
- *accept and manage the risk.*

Project Hazard Inventory

Project Title		Project No.	Project Manager
		Division	Project Safety Co-ordinator

(1) Hazard Ref.	(2) Hazard	(3) Cause	(4) Initial risk level			(5) Design action taken?		(6) Mitigation measures	(7) Final risk level
			Probability	Severity	Risk	Yes	No		
1									
2									
3									
4									
5									

Fig. A2.6 Project risk register for a qualitative approach

A2.4 References

Boothroyd, C. and Emmett, J. (1996). *Risk Management. A Practical Guide for Construction Professionals.* Witherby, London.

CIRIA (2001). *RiskCom Software Tool.*

Clayton, C. R. I. (2001). *Managing Geotechnical Risk – Improving Productivity in UK Building and Construction.*

Godfrey, P. S. (1996). *Control of Risk – A Guide to the Systematic Management of Risk from Construction.* CIRIA, London. CIRIA Special Publication 125.

Health and Safety Executive (1999). *Five Steps to Risk Assessment.* HSE Books, London.

Hydraulics Research Wallingford (2000). *Developing a Risk Communication Tool (RiskCom). Report on Research Methodology.* CIRIA, London. Funders Report, CIRIA Research Project RP591.

Institute of Actuaries (1998). *Risk Analysis and Management for Projects (RAMP).* ICE, Faculty of Actuaries.

Institution of Civil Engineers/Institute of Actuaries (1998). *Risk Analysis and Management for Projects – The Essential Guide to Strategic Analysis and Management of Risk. The RAMP Report.* ICE/IA, London.

Simon, P., Hilson, D. and Newland, K. (1997). *Project Risk Analysis and Management Guide. The PRAM Report.* Association of Project Management, High Wycombe.

Thompson, P. A. and Perry, J. G. (eds) (1998). *Engineering Construction Risks – Implications for Project Clients and Project Managers.* Thomas Telford, London.

Bibliography

In addition to the references at the end of each chapter, the following is further reading.

Adachi, T. (1992). *Simulation of Fenner-Pacher Curve in NATM, Numerical Models in Geomechanics.*

Amstutz, E. (1970). Buckling of pressure shaft and tunnel linings. *Water Power and Dam Construction.* November, pp. 391–399.

Atkinson, I. H. and Potts, D. M. (1977). Stability of shallow circular tunnels in cohesionless soil. *Géotechnique* **27**, No. 2, 203–215.

Attewell, P. B., Yeates, I. and Selby, A. R. (1986). Soil movements induced by tunnelling and their effects on pipelines and structures. Blackie, Glasgow.

Barton, N. R., Grimstad, E., Aas, G., Opsahl, O. A., Bakken, A. and Johansen, E. D. (1992). Norwegian method of tunnelling. *World Tunnelling*, June. The Mining Journal Ltd, London.

Barton, N. R., Grimstad, E. and Palmstrom, A. Design of tunnel support: Sprayed concrete properties, design and application. (eds Austin, S. and Robins, P.).

Barton, N. R., Lien, R. and Lunde, J. (1975). Estimation of support requirements for underground excavations. *ASCE Proceedings 16th Symposium on Rock Mechanics*, September.

Bouvard, A., Colombet, G., Panet, M. and others (2000). *Tunnel Support and Lining.* AFTES, France.

Copperthwaite, W. C. (1906). *Tunnel Shields and the Use of Compressed Air in Subaqueous Works.* Archibald Constable, London.

Davis, E. H., Gunn, M. I., Mair, R. J. and Seneviratne, H. N. (1980). The stability of shallow tunnels and underground openings in cohesive soils. *Géotechnique* **30**, No. 4, 397–416.

Duddeck, H. (1988). Guidelines for the design of tunnels. *Tunnels & Deep Space* **3**, No. 3.

European Federation of Producers and Applicators of Specialist Products for Structures (EFNARC) (1996). *European Specification for Sprayed Concrete.*

Geospatial Engineering Practices Committee of the Institution of Civil Engineering Surveyors. Monitoring Guide being drafted.

Golder Associates and MacLaren Ltd (1976). *Tunnelling Technology: An Appraisal of the State of the Art for Application to Transit Systems.* Ontario Ministry of Transportation and Communications, Toronto.

Golser, J. and Hackl, E. Tunnelling in soft ground with the New Austrian Tunnelling Method (NATM). *9th Int. Conf. on Soil Mechanics and Foundation Eng.*, Tokyo.

Hewett, B. H. M. and Johannesson, S. (1992). *Shield and Compressed Air Tunnelling.* McGraw-Hill, New York.

Holmgren, I. (1993). Design of shotcrete linings in hard rock in *Shotcrete for Underground Support VI – Proc. of Eng. Foundation Conf.*

Holmgren, I. (1993). Thin shotcrete layers subjected to punch loads in *Shotcrete for Ground Support.* Ibid.

Innenhofer, G. and Vigl, L. (1997). Tunnel lining without reinforcement in *Tunnels for People Conf. Proc.* Balkema, Rotterdam.

Jacobson, S. (1974). Buckling of circular rings and cylindrical tubes under external pressure. *Water Power*, December, 400–407.

Ladanyi, B. (1974). Use of the long term strength concept in the determination of ground pressure on tunnels linings. *Proc. of the 3rd Cong. of the International Society of Rock Mechanics*, Denver.

Lauffer, H. (1958). Gebirgsklassifizierung für den Stollenbau. *Geologie und Bauwesen* **24**, No. 1, 46–51.

Leca, E. (1996). Modelling and prediction for bored tunnels in *Geotechnical Aspects of Underground Construction in Soft Ground.* Balkema, Rotterdam.

Lombardi, G. (1973). Dimensioning of tunnel linings. *Tunnels and Tunnelling*, July.

Lunardi, P. (1997). Pretunnel advance system. *Tunnels & Tunnelling International*, October.

Mair, R. J. (1993). Developments in geotechnical engineering research: application to tunnels and deep excavations; Unwin Memorial Lecture 1992. *Proc. of the Instn. of Civ. Eng.*, February, Thomas Telford, London.

Mair, R. J. (1996). Settlement effects of bored tunnels in *Roc. Intl. Symp on Geotechnical Aspects of Underground Construction in Soft Ground.* (eds Mair, R. J. and Taylor, R. N.). Balkema, Rotterdam.

Malmberg, B. (1993). Shotcrete for rock support: a Summary Report on the state of the art in 15 countries. *Tunnelling and Underground Space Technology* **8**, No. 4, Pergamon Press Ltd, Oxford.

Megaw, T. M. and Bartlett, J. V. (1981). *Tunnels: Planning, Design, Construction, vols. 1 & 2*, Ellis Horwood, Chichester.

Method and verification of design method (1988). *Canadian Geotechnical Journal*, No. 25.

Monsees, I. E. and Lorig, L. I. (1993). Design of shotcrete support in the United States in *Shotcrete for Underground Support VI, Proc. of Eng. Foundation Conf.*

Moore, E. T. (ed.) (1989). Tunnels and shafts, Chapter 3. In *Civil Engineering Guidelines for Planning and Designing Hydroelectric Developments*, Vol. 2, Waterways. ASCE, New York.

Moyson, D. (1994). Steel fibre reinforced concrete (SFRC) for tunnel linings: A technical approach in *Tunnelling and Ground Conditions*. Balkema, Rotterdam.

Nanakorn, P. and Horii, H. (1996). A fracture-mechanics-based design method for SFRC tunnel linings. *Tunnelling and Underground Space Technology*.

Panet, M. (1992). *Ouvrages Soutterrains – Conception, Realisation, Entretien*. Presses ponts et chaussées, Paris.

Panet, M. (1995). *Le Calcul des Tunnels par la Methode Convergence-Confinement*. Presses ponts et chaussées, Paris.

Peila, D. (1994). A theoretical study of reinforcement influence on the stability of a tunnel face. *Geotechnical and Geological Engineering* **12**.

Pequignot, C. A. (1963). *Tunnels and Tunnelling*. Hutchinson, London.

Pöttler, R. Fresh shotcrete in tunnelling: Stresses, strength, deformation.

Pöttler, R. (1990). Time-dependent rock-shotcrete interaction: A numerical shortcut. *Computers and Geotechnics*, No. 9.

Powderham, A. J. (1994). An overview of the observational method: Development in cut and cover and bored tunnelling projects. *Géotechnique* **44**, No. 4, 619–636. Thomas Telford, London.

Powell, D. B. and Beveridge, J. P. (1998). Putting the NATM into context. *Tunnels & Tunnelling International*. January, p. 41.

Rabcewicz, L. V. and Golser, J. (1973). Principles of dimensioning the supporting system for the New Austrian Tunnelling Method. *Water Power* **25**.

Sauer, G. *Further Insight into the NATM*. 23rd Sir Julius Wernher Memorial Lecture. Institution of Mining and Metallurgy, London.

Schmidt-Schleicher, H. German guidelines for steel fibre reinforced shotcrete in tunnels with special consideration of design and statical aspects.

Swamy, R. N. and Bahia, H. M. (1995). The effectiveness of steel fibers as shear reinforcement. *Concrete International*. March.

Swoboda, G. and Moussa, A. (1994). Numerical modelling of shotcrete and concrete tunnel linings in *Tunnelling and Ground Conditions*. Balkema, Rotterdam.

Swoboda, G. and Wagner, H. (1993). Design based on numerical modelling a requirement for an economical tunnel construction. *RETC Proceedings*.

Watson, P. (1997). NATM design for soft ground. *World Tunnelling*. November.

Whittaker, B. N. and Frith, R. C. (1990). *Tunnelling: Design, Stability and Construction*. Institution of Mining and Metallurgy.

Wong, R. C. and Kaiser, P. K. Design and performance evaluation of vertical shaft: rational shaft design.

Index

Note: Figures and Tables are indicated by *italic page numbers*

concrete linings
  acid attack, 46
  alkali–silica reaction, 46
  carbonation-induced corrosion of reinforcement, 44
  chemical attack, 44–46
  chloride-induced corrosion of reinforcement, 43–44
  codes and standards for, 48–49
  design and detailing factors, 43–47
  detailing of precast concrete segments, 48
  durability, 41–42
    factors affecting, 41–42, 43–47
  impermeable, 89
  physico-mechanical processes affecting, 47
  protective systems, 47–48
  sulphate attack, 45–46
consolidation, 68
constitutive modelling, 99–100, 110–111
  geotechnical model, 110
  lining model, 111
Construction (Design and Management) Regulations (CDM Regulations), 15, 17, 59
Construction Quality Plan, 142
  requirements, 143
construction sequence
  effect on design analysis, 99
  settlement affected by, 116–117
  sprayed concrete lined tunnels, 82
continua, design methods for, 101, 104–106, 106, 107, 108
continuum analytical models, 101, 104–105
Convergence–Confinement Method (CCM), 101, 104, 105
corrosion
  protection, 12, 42–43
  rates, 40
    factors affecting, 43
costs, capital vs maintenance, 13
cracking effects on concrete, 47
critical strain(s)
  definition for rock masses, 92
  in ground, 92–95
Crossrail tunnels, design life, 12

data acquisition and management, I & M system, 132–134
data-loggers, 132
definitions, 3–4
deformation(s), design considerations, 64
design, 59–97
  characteristics, 61
  definition, 3
design analysis, purpose, 98
design considerations, 63–75
  choice of lining systems, 73–75
  effects of ground improvement or groundwater control, 69–71
  engineering design process, 61–63
  excavation methods and, 71–73
  fundamental concepts, 60
  ground improvement, 69
  ground/support interaction, 63–65
  groundwater, 68
  pre-support, 68–69
  time-related behaviour, 65–68
design development statements, 139
design life, 12–13, 40, 153
design methods, 100–113
  advances in numerical analyses, 111–112
  'closed-form' analytical methods, 101, 104–106
  constitutive modelling, 110–111
  discretisation in, 108–109
  empirical methods, 101, 102–104
  interpretation, 100
  modelling construction processes in, 109–110

modelling geometry for, 108
  numerical methods, 101, 106–108
  recommendations on, 113
  theoretical basis, 100
  validation of models, 111
design process, 5–7
  concept stage, 5
  outputs, 139
Design Quality Plan, 139, 142
designer
  health and safety responsibilities, 18–19, 59
    challenging accepted approaches, 19
    information provision, 18
    record keeping, 18–19
    risk avoidance, 18, 59
  meaning of term, 59, 138n
desk study, 22
detailing
  gaskets, 57
  precast concrete segments, 48
  sprayed concrete lined tunnels, 82
dewatering, 37
  effect on lining design, 70
discontinua, design methods for, 101, 102–104, 106, 107–108
discrete element (DE) method, 101, 106, 107
distortion of linings, 92
  recommended ratios for various soil types, 92
drained (water management) systems, 54–55
drill-and-blast technique, 73
driven tunnel, definition, 4
dry caisson techniques, shafts installed using, 85
durability considerations, 11, 12, 40–42
  design and specification for, 42–49
  factors affecting, 41, 162
  and lining type, 40–42
  and tunnel use, 40
durable lining, definition, 40
dynamic compaction, ground improvement by, 37, 71

earth pressure balance tunnel-boring machines (EPB TBMs), 61, 66–67, 68
  effect on lining design, 161
earth pressure monitoring, 126
electrolevels, 126
empirical (design) methods, 101, 102–104
  disadvantages, 103
  see also observational method
engineering design process, definition, 4
environmental considerations, 8, 13
  external environment, 13
  internal environment, 13
EPDM compression gaskets, 56, 78
Eurocode 1, 52
Eurocode 2, 48
European Code for Concrete, 40
European Directives
  Temporary or Mobile Construction Sites Directive, 16
    duties defined, 16
    shortcomings listed, 16
    UK legislation implementing, 16–17
European standards, concrete structures, 48
excavation (ground investigation) techniques, 23
excavation method, effects on lining design, 71–73
explosive spalling due to fire, 51
  control of, 49

face sealing, ground improvement by, 69
factory-produced linings
  documentation for, 140
  lifting and handling of, 75, 76
  load test for, 140
  proof ring build, 140

quality management issues, 139–140
  see also segmental linings
Federal Highway Administration (FHWA), on
    instrumentation and monitoring, 134
fibre optical strain instruments, *127*
field investigation and testing methods, 22, *23*
finite difference (FD) method, *101*, 106, 107, 108
finite element (FE) method, *101*, 106, 107, 108
fire
  concrete behaviour in, 50–51
  lining material behaviour, 50–51
  and material properties, *51*
  types, 50
fire protection measures, 53
fire repair, 53–54
fire resistance, 12, 43, 49–54
  codes and standards covering, 52–53
  effects of tunnel type and shape, 50
flexibility of linings, 90–92
  monitoring of deformation, 141–142
forepoling, 68
form of contract, 14
freeze–thaw attack on concrete, 47
functional requirements for linings, 8
funding, and form of contract, 14

gas permeability, 13
gaskets, 56–57, 78
  design and detailing of, 57
  injectable, 57
Gaussian model, for ground movements, 115–116
geodetic surveying targets, *125*, 165
Geological Strength Index (GSI), 27, 103
geophysical ground investigation techniques, *23*
Geotechnical Baseline Report (GBR), 29, 37
  as part of Contract, 37–38
  primary purpose, 38
geotechnical characterisation, 20–38
geotechnical design parameters, *32*
  applications, 31, *32*
  identifying patterns, 33–34
  maximum and minimum bound, 34–35
  obtaining relevant information, 31, 33
  range and certainty, 31, 33–36
  sensitivity analysis, 35–36
Geotechnical Design Summary Report (GDSR), 29
Geotechnical Factual Report (GFR), 28, 37
Geotechnical Interpretative Report (GIR), 28–29, 37,
    38
GEotechnical Monitoring INformation Interchange
    (GEMINI), 123
geotechnical parameters
  Channel Tunnel, *152*
  laboratory test methods for, 24–25
  London Heathrow Express T4 platform tunnels, *145*,
    146
  required for tunnel lining design, 31–36
  scatter of test results, 20
geotextiles, in waterproofing systems, 55
Great Belt railway tunnels, 159–163
  geology, 159–160
  geotechnical parameters, *160*
  linings
    cross-section, *159*
    design, 161
    durability, 162
    joint details, 162, *163*
    reinforcement in, 162
    watertightness, 163
  route, *159*
'greenfield' check of instrumentation, 129
grey cast iron linings, durability, 12, 40
grit blasting, problems on coating, 42–43
ground conditions, reference, 20, 37

ground freezing, 37, 70–71
  monitoring of, 124
ground improvement methods, 36–37, 69
  compaction, 36
  dynamic compaction, 37
  effects on lining design, 69–71
  jet grouting, 36, 69
  permeation grouting, 36
  vibro-replacement, 36
ground investigation, 20–25
  factors in selection of methods and scope, 25
  'foreseeing the unforeseeable' exercise, 30–31
  interpretation of data, 28–29
  process, 20–22
ground loss
  definition, 4
  modelling of, 115
ground model, 20
  instrumentation and, 129
  for London Heathrow Express project, 144–145
ground movements
  advance settlement, 116
  characterisation of, 115
  counteracted by grouting, 70, 118–120
  effects, 117–118
  Gaussian model for, 115–116
  prediction of, 115–117
  three-dimensional models, 116
Ground Reaction Curves (GRCs), 64–65
ground response, 90
ground/support interaction
  in design considerations, 63–65
  modelling of, 64–65, 117
groundwater
  behaviour, 29–30
  changes in water table, 36, 70
    effects on ground parameters, 36
  control methods, 37
    exclusion methods, 37
    ground freezing, 37
    low-pressure compressed air, 37
  design considerations, 30, 68
  identification in soils and rocks, 27–28
  pollution of, 13
grouting
  of cast in situ linings, 83
  ground improvement by, 36, 69, 70
  leakage prevention by, 57–58
  of segmental tunnels, 78–79
Grozier Tunnel (Switzerland), 55

hand tunnelling, 72–73
hard ground
  definition, 4, 29
  ground investigation techniques for, *23*
  groundwater effects, 27
  transition to soft ground, 29
hazard, definition, 4
hazards identification, 11
health and safety
  designers' responsibilities, 9, 18–19
  legislation, 14–15, 59
Heathrow... *see* London Heathrow...
Hoek–Brown yield criteria, 161
human error, 100
'hybrid ring' concept, 158
hydrocarbon fires, 50, 52
hydrophilic seals, 56, 78
hypothetical modulus of elasticity (HME), 111

impact
  meaning of term, 173
  scoring, *172*
impact effects on concrete, 47

impermeable linings, 89
inclination, monitoring of, *127*
inclinometers, *126*
information provision, health and safety aspects of
  design, 18
injectable gaskets and seals, 57
instrumentation and monitoring (I & M), 122–136
  in CTRL North Downs Tunnel, 132, 134, 164–165
  data acquisition and management, 132–134
  data format, 130–131
  data management for, 130
  designer's checklist, 129–131
  examples in case histories, 134–135
  existing guidance on, 123
  frequency of readings, 130
  'greenfield' check, 129, 131
  limitations, 129
  and lining design, 123–131
  in London Heathrow Express project, 147
  management of third-party issues, 131–132
  methodology, 130
  necessary instruments only, 129–130
  in observational method, 128–129, 133
  purposes, 122, 123–124
  real-time data acquisition, 130
  reliability, 130
  resolution, 129
  responsibility for, 130
  tunnelling data included, 130
  typical applications, *125–128*
  value of, 122
International Tunnelling Association (ITA)
  design guidelines, 1, 54, 60
  'model' tunnel design process, 7, 98, *99*
Interpretative Geotechnical Report, 28–29, 37, 38, 160
invert closure, 66–67, 80
iron *see* grey cast iron . . .; spheroidal graphite iron
  (SGI) . . .
ISO 834 (on fire resistance tests), 52

jet grouting, ground improvement by, 36, 69
Joint (BTS/ABI) Code of Practice for Risk
  Management of Tunnel Works in the UK, 17–18
  contents, 17–18
  scope, 17, 38
joint movement, monitoring of, *128*
joints, in precast concrete segmental linings, 77, *155*,
  162, *163*
Jubilee Line Extension (JLE) tunnels, 12, 90, 112, 123
junctions
  design considerations, 86–87
    constructability, 87
    ground stability, 87
    structural stability, 86–87
  sprayed concrete lining for, 82

key performance indicators (KPIs), 90
Kielder experimental tunnel, 112

laboratory test methods, for geotechnical parameters,
  24–25
lateral displacement, monitoring of, *127*
launch chambers, design considerations, 87–88
leakage in lining, monitoring of, *128*
leakage-prevention grouting, 57–58
likelihood
  meaning of term, 173
  scoring, *173*
limit-equilibrium analytical methods, 105
lining, definition, 4
lining systems
  cast in situ linings, 75, 83
  choice of, 73–75
  segmental linings, 73–74, 75–79

sprayed concrete linings, 74–75, 79–83
liquid-level settlement gauges, *125*
loading considerations, 9–11
  external loading, 9, *10*
  internal loading, 9, *10*, 11
London Clay, geotechnical parameters, *145*, *146*
London Heathrow Airport
  Cargo Transfer Tunnel, 38
  Terminal 4, 144
  Terminal 5, 56
London Heathrow Express, 144–148
  design analysis, 145–147
  design approach, 144
  excavation sequence, *147*
  geotechnical parameters, 144, *145*
  ground model for, 144–145
  instrumentation and monitoring, 147
  key performance indicators, 90
  lining details, 147
  platform tunnel geometry, *148*
  project background, 144
  trial tunnels, 112, 146
London Underground Jubilee Line Extension tunnels,
  12, 90, 112, 123
loosening concept, 65
Lyon Metro tunnels, 55

machine tunnelling
  in rock, 72
  in soft ground, 71–72
  see also tunnel boring machines (TBMs)
maintenance costs, capital costs vs, 13
Manufacture Quality Plan, 139, 142
manufactured linings *see* factory-produced linings
masonry linings, 83
materials
  environmental effects, 13
  selection of, factors affecting, 11–12
measurement method, and risk apportionment, 14
Melbourne City Tunnel, 54–55
membranes, waterproofing, 54–56
metal linings, design and detailing factors, 42–43
modulus ratio
  bending moment varying with, *91*
  definition, 91
  hoop thrust varying with, *91*
monitoring
  lining deformation, 141–142
  real-time, 130
  surface settlement, 142–143
  trigger values for, 133–134
  of tunnel lining behaviour, 124
  see also instrumentation and monitoring
multiple tunnels, design considerations, 88

neoprene gaskets, 56
neural networks, 111–112
New Austrian Tunnelling Method (NATM), 61, 81
noise, effects, 13
non-circular tunnels, effect of fire, 50
North Downs Tunnel *see* Channel Tunnel Rail Link
  (CTRL) North Downs Tunnel
Norwegian criterion, 88–89
numerical modelling, *101*, 106–108
  advances in numerical analyses, 111–112
  ground movements, 117
  validation of models, 111

objectives of Guide, 1–2
observational method (OM), 103–104
  definition of term, 103–104, 129
  design management and, 61
  instrumentation and monitoring as part of, 128–129,
    133

Tunnel lining design guide. Thomas Telford, London, 2004